Co-Operativization Liberalization and Dairy Industry in India

Co-Operativization Liberalization and Dairy Industry in India

Dilip Shah

ABD PUBLISHERS
Jaipur ■ New Delhi

ABD PUBLISHERS

Regd. Off. "Bony Residency", Gate No. 2
Opp. Tilak Public School, Vishweswariya Nagar,
Gopalpura Road, Jaipur - 302018, India
Ph: 0141 - 2761280, Telefax: 2761381
Email: abdpublishersjpr@gmail.com

Delhi Office: 102, 1st Floor, Satyam House,
4327/3, Ansari Road, Darya Ganj,
New Delhi - 110002, India
Ph: 011-45652440

visit us: www.oxfordbookcompany.com

Distributed By
KEYA INTERNATIONAL PRESS LLP
E-mail: keyainternationalpressllp@gmail.com

Published in India, 2025

Price: ₹ 2400/-

ISBN: 9788119343652

Typeset by Shivangi Computer, Jaipur

Printed and bound in India by Thomson Press Ltd.,
New Delhi

Dedicated to

USHA

my

beloved wife

FOREWORD

"Is it really cooperatives (have) performed so poorly that we can't hope for any effective role or cooperativization in India today? "asks Prof. D.R.Shah in the book while discussing in the introduction, the future of cooperatives in the country inthe context of the New Economic Policy (NEP). Asking further, "Are these issues possible to resolve?", Prof Shah discusses future prospect after reviewing the cooperative development scenario and makes out a case for more vigorous efforts and a dynamic National cooperative Policy to facilitate a challenging role for the cooperatives to counter the inroads the MNCs are making into the areas served by cooperatives after sustained efforts during the past few decades.

Prof. Shah's present publication is a collection of papers on New Economic Policy initiatives of the government and its impact on future cooperative development; the need for fresh liberalization legislation in the context of economic liberalization and several papers dealing with dairy cooperative industry and its problems relating to NEP,dairy, export performance, entry of private dairies in Mehsana and Baroda and cooperative dairy development in the year 2000 A.D. paper also deals with cooperative and rural development in the context NEP. Dr. Shah in his closing chapter has given latest scenerio of the liberalised Dairy Industrial Policy and their Policy Implications in next melleniuam and thus, has up date the book.

While analysing the causes for cooperative failures. Prof.Shah has referred to Shah's advocacy of evolving rules within cooperatives to suit local conditions rather than rigidly adhering to international cooperative principles. He has referred to Prof. Dandekar's statement on overdues which, he feels, are largely due to the collective responsibility of all members in a cooperative. It seems some of the writers feel that the dogmatic approach to observance of basic cooperative principles can lead to cooperative failures. It is an age-old controvers between the 'member based' and the value based' approaches, Prof. Shah has also handled very effectively the arguments for and against the blue print approach and the 'green house approach' while discussing the question of replicating the Anand Pattern or similar success models in the cooperative dairy sector. The Author has also attempted to analyse the background against which the Government had to embark upon the new industrial policy followed by NEP and has suggested that pressures from the IMF, the world Bank, GATT and the USA, have to some extent, forced the Government to initiate measures in this direction. The justifioation put forward by the government in support of NEP. according to Prof. Shah, were : competition, investment, growth of output, technology promotion and export promotion.

The papers in the book are largely devoted to the problem of the dairy scotor and Prof. Shah has given very useful information and data on the influence of the private dairies in the rural area of Mehsana and Baroda. He has suggested that the private dairies should be controled to handle fresh milk and not be allowed to concentrate on finished products. For meeting the challenge in the export market, the Author has advocated adaption of Anand Pattern, change of out-dated cooperative laws, greater role to NDDB, improvement of quality and a balanced and not too liberal role for the privets sector.

Prof. Shah rightly points out to the need of improvement of the quality of cooperative services to their members, He had also advocated the formation of cooperative companies to bring about greater flexibility in cooperative dairy Institutions apportions. He feel that if the cooperatives adopt high technology and cut their operating costs, they have greater chances to stand up the emerging competition. In his paper dealing with cooperatives and rural development in the context of NEP, Prof. Shah suggests the Middle path for future cooperative strategy which "capable of combining the efficiency and Productivity of the market-friendly liberalization with socialist concern of welfare and productivity.

He has also endorsed the statement of the Cooperative Initiative Panel (CIP) regarding emphasis by cooperatives on action of awareness among members and on the education of the people involved in cooperative and also of the people in the surrounding areas.

Prof. Shah's laudable attempt to bring about a discussion on the various facets of NEP and their implications is a welcome addition to coperative literature and students of cooperation would find it useful and informative. Prof. Shah may consider a more intensified effort, in a more logical sequence, on the dairy cooperatives alone as and when a second edition of the books is envisaged.

M.V. Madane

Poone

PREFACE

It gives me a great pleasure to welcome the well researched work of Dr Dilip Shah in the literature of Co-operative Studies and Co-operative Management. Indian economy is basically rural, and agriculture and agro-based industry constitute the back bone of rural economy.

Indian economy is highly influenced by the performance of rural economy which contributes a lion's share both in terms of its significant share in the nation's gross domestic product as well as number of players government of India has, therefore put rural development on its list of economic priorities through numerous schemes. Government of Gujarat has made rural development its focal point for economic resurgence by promoting "Gokul Gram Yojana" in a big way. To realise the goals of governmental efforts, the need to strengthen the co-operative movement can not be overemphasized. The need to study the co-operative sector and the various sub-sectors within it, to make it competitive with other sectors by improving its performance through effective participative management practices is the need of the hour.

In this context, the present work of Dr. Shah provides a fresh insight into the latest scenario of Co-operative sector in general and dairy co-operative sector in particular. This study will be useful to all those who wish to review the recent theoretical literature and empirical studies, which are well documented in the book. This study also records `ground realities' which must be changed through appropriate, timely policy devices including National Co-operative policy and Co-operative legislative reforms. This book is of much significance for those who are interested in Co-operative Dairy Development because of Dr. Shah's analytical observations of the changing scenario of Dairy Industry right from pre-liberalization period to the present time when we are approaching the next millennium.

I recommend this book to all those who may be interested in the Co-operative Development and Co-operative Management studies.

PREM SHARDA
VICE-CHANCELLOR
SOUTH GUJARAT UNIVERSITY
SURAT

ACKNOWLEDGEMENT

We sinceely acknowledge the kind permission given by Institutions and agencies who have published our papers which are included in this volume. We also take the opportunity of the funding agencies who have supported our research projects and also allowed us to re-publish the report of the sponsored research. Specifically, we are grateful to Indian Society of studies in Co-operation, Vaikunth Mehtra Institute of Co-operative Management, Editors of Co-operative Perspective Financial express and Discovering Co-operative and Rural Co.operatives and development News Bulletin of N.D.D.B. and Insitute of Rural Management etc.

Once again we gratefuly acknowledge the support of all those who have help to contribute in acadamic persuite.

INTRODUCTION

Co-operative sector is a 'THIRD SECTOR' of Indian Economy after Private and Public Sector in India. It is world's second largest co-operative sector with great diversification. Though its major achievement and success in Agriculture and Rural development are confined to selective regions and economic activities and yet it would not be wise to gloss over the potentiality of building new rural India based on powerful co-operative structure in next millennium. The liberalization of Indian economy in general and co-operative sector in particular is a challenge and opportunity both. The co-operative sector will have to be efficient to work as safety net for rural poor producers and also consumers as market failure to serve these class is very usual. The success stories and models in co-operatives are the great sign post of such a successful sector provided we develop an environment to replicate the success stories as regular character of Indian co-operative sector.

Present set of the papers written at different times in recent period is an attempt to develop a positive thinking and outlook based on the research studies conducted by different scholars. A couple of research papers are also based on our research studies. Almost all these papers are published in relevant Journals. I am extremely encouraged by ' Foreword' to our book by eminent scholar and expert of International level Dr. Madane, who is also the Chairman of Indian Society of studies in co-operation. I am very grateful for his kind favor to me. I am equally grateful to our Vice Chancellor Prem Sharda for extending his encouragement to me by way of writing preface to my book. Our efforts and painful efforts of the publisher will be rewarded if these papers will be found interesting and useful for those who are concern with co-operative studies.

Surat DILIP SHAH

CONTENTS

1

CO-OPERATIVIZATION IN INDIA

Need For A Fresh And Positive Approach, Initiative And Actions

The New Economic Policy aiming at adjustments and restructuring of Indian economy provides no place and reference to its third sector of economy except a negative measure of deregulating and delicensing policy and thus opening the flood gates of competition throwing co-operatives in to venerable position due to very defective, discriminative and hasty style and approach in dealing with the process of "opening up" all sectors. This provides an evidence of skeptism on the part of government with tendency of the government to undermine the role of co-operatives is further evident from the document of VIIth plan where even traditional ritual of providing a separate chapter on co-operatives have been withdrawn as if co-operatives have nothing to do with development plan. Under this shadow of the gloomy environment for the promotion of co-operatives, we intend to focus on the basic questions with fresh evidence that, Is it really co-operatives performed so poorly that we can't hope for any effective role of co-operativization in India today ? What are the most basic issues which come in the way of powerful practice of co-operativization in India ? Are these Issues possible to resolve ? How? These are the questions, we propose to deliberate in this paper.

QUICK REVIEW OF CO-OPERATIVE DEVELOPMENT SCENARIO:

Before we quickly draw an emerging picture of the co-operative development scenario; we may look back over history of co-operative movement particularly during the planned period to record the objective of co-operativization in India. (Prior to Independence, it began as an activity exclusively patronage projected and promoted by Government as a relief programme for credit and as result, co-operatives in India were largely identified as government credit agency until planned period, however, First plan aimed at co-operatives to develop as a "Third Sector" of Indian Economy aiming at balancing private sector and Public sector of the economy (1) Incidentally, there were visionaries like D.R. Gadgil who went further and even looked forward to develop "Co-operative common wealth" (2) However, as we shall see presently, every plan assigned some role to co-operative sector except eighthhplan. Second five year plan expected the co-operatives to play a vital role on agricultural and Industrial fronts. Co-operative envisaged to play a crucial role in structural transformation of agrarian economy by co-operative farming as well as by promoting consumers goods Industries on small scale and village/ cottage Industries within co-operative stores. In euphoria of new strategy of agricultural development and consequent green revolution, co-operatives were made carriers of Inputs along with credit supply and provider of marketing linkages during IVth and Vth plan. In sixth plan and seventh plan, we thought to develop co-operatives for development of rural poor to redress the distributional distance of development which started becoming obvious by this time. Thus, what we find a clear evidence of the shifting and confusing co-operative policy. It is notable that co-operatives did play whatever role assigned to it quite successfully, but despite that, we lacked an integrated, stable and powerful co-operative sector. This state of affairs could be attributed to absence of National co-operative Policy. (However, the following quick review to provide us an evidence that there are enough reasons to hang our hope for the unignorable, effective and powerful co-operativization in India). We may quickly record and analyse the main dimensions of co-operative development scenario in terms of its size, structure and spread in India.

SIZE OF THE CO-OPERATIVE SECTOR

Recent report of International Co-operative Alliance records that 72 country member of I.C.A. had 7,40,000 co-operatives in the world out of which 35% co-operatives were working in Agricultural co-operatives sector, 28% in credit sector and 0.95% were confined to consumers co-operatives. Thus, 72% of co-operatives related its activity to these three major area of economic activities. (3) India's place in the world co-operative sector was found to be second next to China in terms of its membership. The size of the co-operative sector could be judged from the fact that it has around 3.5 lakh co-operatives spreaded in the villages of India with coverage of 145 Million members and working capital of Rs.48,000 crores, (4) Infact, next to 'Panchayat' co-operative is the only most known institution since ages among villages.

STRUCTURAL TRANSFORMATION OF CO-OPERATIVE SECTOR

It is quite known fact that government initiated co-operative movement in early period of 19th century by developing co-operative as a Agricultural credit sector.(5) This historical pattern of dominating credit co-operatives has been retained almost until 1960s. In the last decade from the level of just Rs. 23 crores worth credit disbursement in 1980. It reported to disbursed as much Rs. 5558 crores worth credit in 1990/91. Further, the real significance achieved by co-operatives could be seen in its effective participation in Institutional finance in rural sector. This could be found from the latest report of Reserve Bank of India which records that the share of rural households debt, which was just 3% from co-operative credit sources have significantly increased to 35%.[6] Moreover, remarkable and better achievement have been recorded by co-operative sector as compared to Commercial Bank sector in extending finance to weaker sections.[7] This study has evidently proved that "this weaker section class does not seem to be suffering from any specific discrimination against them in terms of allocation of resources".[8]

However, the 'credit sector' has retained its significance, the undeniable and remarkable structural changes in co-operative activities through qualitative and quantitative diversification have recorded particularly after 1960's. (A perceptible shift away from the traditional credit co-operatives to non-traditional, non-credit co-operative got initiated from 1954 and by 1989/90, the major transformation is evident from the fact that non-credit co-operatives constituted more than 60% of the number of industries in the co-operative sector of India and the proportion of non-credit co-operative in the co-operative sector in India as a whole almost doubled between 1961/89.[9]) As per the latest report of NABARD, there are as many as 197820 non-credit co-operative with membership of 415050, 0 and working capital of 17473 crores. An extent of diversification could be judged by the fact that as many as 38 types of economic activities are covered under the co-operatives such as Agricultural Marketing, processing, Milk supply, fisheries, poultries, weavers and industrial co-operatives, dairy co-operatives sugar mills handloom mills, khadi and village Industries co-operatives, construction labour co-operatives and Forest labourers co-operatives. Input supply factories and co-operative fertilized plants and cattle feed and poultry feed plants edible oil mills, tree growers and groundnut growers co-operatives, salt producers and fruits and vegetable growers co-operatives, tube-well co-operatives and water co-operatives etc.

The progressive process of transformation from a traditional credit co-operative structure to non-credit co-operative structurel has exhibited some other remarkable directional which are brilliantly brought out by S.S. Puri[10]. For example, the number of primary agricultural co-operative has drastically gone down, the number of such co-operative stood at about 212000, at the end of 1961, which declined to 87000 in 1989. It has been attributed to as the outcome of reorganization and amalgamation of primary agricultural credit societies to achieve 'financial viability'. Another point of direction is the emergence and growth of non-credit co-operatives in non agricultural sector or we may say in urban areas particularly in the area of consumers co-operative and Input co-operative and housing and industrial co-operative in general and small and village industries handloom and handicrafts in particular. Similarly, in the rural sector itself manifestations and its intensity alongwith implications for the development are also noteble. We have tried to analyse the available latest interstate facts and data relating to co-operativization in the states of India. We have attempted to look into the question of interstate comparative progress and spread of co-operatives with reference to broadly available indicators with reference to credit co-operatives, non-credit co-operatives, Agricultural primary credit co-operatives and major non-credit co-operatives as recorded by NABARD in its latest edition of Important Items of data relating to co-operatives movement, some of the interesting points of the observations are briefly stated as follows:-

(A) Spread of credit co-operative sector

We may examine the concentration of CCS (Credit Co-operative Sector) in terms of the three variables namely, Number of co-operative society, members of the societies and paid up capital of societies. Some of the striking observations could be stated as follows :-

(I) In terms of all the three variables Maharashtra stands first, however, extent of concentration of CCS in terms of society is higher i.e. 28% while, the share of Maharashtra in membership is almost half of its % share in societies thus, indicate the milder concentration.

(II) Unlike first ranking, in second rank, West Bengal comes next to Maharashtra, but in terms of membership, it is Uttar Pradesh and it is in terms of paid up capital again Uttar Pradesh.

(III) If we take into account, aggregate share of first five states; In most cases more than 70% to 80% share is contributed in these states; thus, it becomes obvious that there is a considerable regional concentration.

(B) Spread of Non-Cooperative Sector

(1) The regional concentration is sharper and specific in the case of non-cooperative sector, despite this fact, Maharashtra as the first ranked state in non-cooperative sector also continued though with

lesser intensity, when compared with different variables such as membership and paid up capital. Assam retains its rank.

(2) The second and third ranking goes on changing according to measure of concentration. For example, when we compare regional spread with reference to number of society we find Gujarat stands next to Maharashtra but in terms of membership we find Assam next to Maharashtra and also in case of paid up capital. Assam retains its rank.

(3) As usual when, we aggregate the share of 5 major states, we find, more than 75% to 80% concentration among thoese states.

(C) Structures of Non-credit co-operative sector

It is very interesting that though in most of the cases and in aggregate Maharashtra stand Ist, specific dominances is visible in case of almost all Non-cooperatives. sugar, processing, poultry and fisheries and construction etc. In general marketing and dairying Gujarat ranks first, but in forest labour co-operative, Madhya Pradesh stands first while consumers co-operatives are highest in Kerala. Thus, one could see the non-cooperative sector has spreaded more regionally depending upon the structure of natural resources and economic activities of the regions.

Thus, the quick review of co-operatives in India leads to make the following observations regarding the status of co-operatives in India.

(1) Despite changing policy of focus, the co-operative sector in India has been developed in a size which is third in the world.

(2) Though it initiated as primary credit service organization, by now it has achieved notable structural transformation in the form of cooperativization in non-credit areas of production and services both.

(3) Diversification of cooperativization has touched upon more than 30 areas of economic activities.

(4) The locational significance has shifted to urban area and now it is not necessarily only rural based activity.

(5) Some co-operatives have also achieved remarkable success in terms of their qualitative improvement as well as participation in the economy and development of weaker sections.

(6) The unique characteristics of regional concentration are also retained which are found interesting variations in terms of region, commodity and class of activities. Thus, cooperativization in India exhibits ray of hope for the future.

CRITICAL QUESTIONS CONCERNING CO-OPERATIVIZATION IN INDIA

We have reviewed an extent and nature of co-operativization in India. We shall now discuss some of the major crucial questions which are capable of explaining very largely the current status of cooperativization in India. The focus on the rightly identified issues is very important to chart out the pattern of initiatives and actions needed to revitalize the co-operatives so that a vibrant sector of the economy could be emerged. We have very briefly discussed some of the major issues in the following seven specific dimensions touching the co-operative development in India today.

1. What should be the OBJECTIVE of the co-operativese ?

This is an old question attached with formal evaluation. Generally, the objective of the existence of co-operatives is stated in terms of the implementation of the Principles of Co-operatives however, while there is no denial of the principles of co-operative should be followed in the right sense, there are now interesting arguments made by scholars that some of the basic difficulties of survival of co-operatives come very much from dogmatic operation of such principles. (For example, Dandekar once made the statement that basic problem of 'overdues' is very largely a consequence of the principles of co-operatives also, i.e. For instance principle of the collective mutual responsibility in recovery). If society has overdues, even the members who have no overdues are refused credit because of the principle of mutual responsibility, because of this principle, those who have to repay do not repay for that if the society goes into overdue, they may not get fresh loan. As they donot repay, the society goes into overdues and their property comes true. Similarly, B. Venkatappaih, draws our attention to the dilemmas arising out of voluntarism of co-operation and compulsion of planning.[13] Planning involves a degree of compulsion for the imposition of a pattern of the process of growth and development, cooperation shows the economic value of planning but as a voluntary form of organization, it is based on an ideology which is the antithesis of compulsion. This perhaps is the most fundamental of all the dilemmas.[14] However, recent study of T. Shah has done very remarkable analysis at how principles related and not related to the development of viable and vibrant institutions.[15] Infact, he has analysed the cases of successful co-operative where the principles of co-operatives are violated for development of viable co-operative institutions.

In fact, T. Shah has made[16] a bold statement after doing a study of successful and failed co-operatives that, there is a strong rational basis for member organisations co-operatives to emerge and grow is evident in the fact that they come up "Swayambhoo" and these co-operatives evolves themselves internal rules and principles which have often little to do with International or other principles for people to co-operative with each other .... utility to the members is only basis for co-operative to emerge and sustain and the normative rules that might have used for co-operatives are those demonstrated and hence its ability to achieve purpose important to members better than alternative organizations.[17] These observations raised important question relating to role of normative values, principles of co-operatives and need for review to revise the principles themselves ?

2. What should be Indicators of Performance ?

While a lot of debate has taken place with regard to success and failure of co-operatives, much less academic studies have been devoted to standard, strong scientific methodology of measuring performance. Recent studies do provide some guidelines but all acceptable, standard performance system is yet to be evolved. We illustrate some alternative performance indicators with their limitations. A celebrated study of Baviskar-Atwood measures the performance of co-operative in Western India in terms of (a) quantity, variety and durability of co-operatives (b) spontaneity (c) managerial control with members (d) viability in terms of profitability.[18] While its overall coverage is comprehensive, it suffers from some conceptual problem arising out of its practice, i.e. profitability and managerial control. Tushaar Shah has suggested another more reliable and real criteria. The list includes (1) Financial viability, (2) Basic condition of self-governance, (3) Adherence to the basic principles of co-operatives such as voluntary membership, democratic control and patronaged line surplus, (4) growth trends in membership and business, (5) Centrality to the economies of their membership and (6) ('Downward Rate net' indicating a state of affairs suggesting a degree of stagnation, apathy and irrelevance a syndrome which regarded as 'Downward Rate net'[19]) even these criteria also suffer from certain kind of subjectivity, difficulty of aggregation, contradiction with objectives etc. NIRD experts Mohanan and others in their CIRDAP sponsored study of development of monitoring and evaluation list out as much as 11 organizational, 14 operational and 8 member participation indicators[20]. The scheme seems to be exhaustive and effective but the status of data base in co-operative sector as it is today, perhaps would not allow itself to get implemented extensively. The problems of priorities, synthesis of criteria, aggregation of norms and reconciliation contrasting objectives, appropriateness to fixation of weightage, and required time series data etc. are the problems which are yet to be satisfactorily resolved. Thus, to design appropriate performance sheets to monitor and evaluate co-operatives is a vital issue deserving a more serious academic perusal.

3. **Why Co-operatives become successful in particular Region and in the case of specific commodities ?**

The success stories of co-operatives in Gujarat and Maharashtra with reference to Milk and Sugar have provocated wider interest to analyse the 'Factors' behind the success. This kind of exercise is very valuable to design an appropriate policy to replicate the successful co-operatives in other parts. One most celebrated and scientific effort in this direction is that of Baviskar and Atwood[21]. According to Baviskar-Atwood thesis. there are three basic factors : (1) Social structure, Caste system and dominant role played by Maratha in Maharashtra and Patidar in Gujarat. They had maratha and Patidar for co-operative development in their area because these communities are progressive, Development oriented, advanced in business, politics and enterprise etc. (2) Stable alliance between large and small farmers for fully harassing economies of scale of those plants which are highly capital intensive for processing perishable commodities like sugar and milk etc. The plants are erected by large farmers and raw material is produced by small and marginal producers, therefore, a stable alliance is favourable to large producers. (3) Strong sustainable leadership : This is an outcome of members controlled co-operative where co-operative leaders could repeat their power benefits which have encouraged the growth of powerful leadership. Thus, the secret of the success of vibrant co-operatives in West India attempted to explain in terms of their society economy and policy.

However, this kind of theory of performance has many catches, some of the points could be raised as follows :

[1] As Tushaar Shah has rightly pointed out a theory of high performance must be automatically a theory of low performance as well. Unfortunately Baviskar-Atwood theory do not explain the failures of co-operatives in the some areas of Gujarat and in other cases[23].

[2] The caste dominated explanation shows that where such a unique castes donot exist co-operative can have no scope, and where there are similar castes comparable to Maratha and Patidars, co-operative must be a strong organization, Reddy in A.P., Jats in Haryana and Sardarji in Punjab are such instance and yet records of co-operative not encouraging.

[3] Similarly, there are co-operative is which havedeveloped without any kind of alliance between small and large farmers.

[4] There are co-operatives which have become successful and could not be explained by those factors but some other factors such as demonstration effect and development funds.

Thus, one would observe that Baviskar-Atwood explanation is not adequate if not incorrect Alternatively, Analysing the cases of success, on the basis of IRMA studies, Tussar Shah well summarize the different factors such as regional culture, poor leadership and management bureaucratic and political interference etc as partial explanation, however, after carefully analying cases of success and failures, these study[24] brings out the following conditions which could be attributed to success.

1. Strnog drive to success and strong collective will for the success.
2. Member first model over value-first model.
3. Service to members : 'utility' for members to earn their patronages, a real strength and source of success.
4. Social commitments of co-operatives.
5. Robust design concept.

Each one have been explained in detail[25] however, relevance of these factors is required to be analysed with greater coverage and details. It is interesting that Misra-Shah study[26] developed a new conceptional frame work to analyse the determinants- the success in terms of linkage dependent and linkage independent co-operatives and then relate with reference to Five 'sub-systems' related to their performance namely (a) Government structure (b) critical linkages (c) patronage sub-system and (d) Micro environment. The sets of factors which were identified as sources of success were namely (i) Individual related factors (ii) Concept specific factors (iii) Macro-environment and (iv) design factors. The co-operatives which were found to under performed have been very successfully explained by Misra-Shah in terms of these factors[27]. Thus, it is very much required on our part to develop our correct understanding regarding the chemistry of those elements which generate success and those are generating failures. Though this exercise is continuous process, IRMA studies have done the major task of defining "Relevant conditions of successful co-operatives."

4. How to create successful co-operatives ?

The successful dairy development based On Anand pattern dairy co-operative model originated a National project of replication of Anand pattern under operation flood was regarded as one answer to the question how to create a successful co-operative but it has generated a lot of debate on the basis of the performance of APC in India.[28] However, a recent contribution of Golden Hydens has given an interesting theoretical account of this issue in different way, where he has advocated 'green house' approach instead of 'Blue print approach'.[29] On the basis of the experience of African bureaucratics he deliberates about the blue print and green house approach. 'Blue-print' approach implies, according to Hyden a notion that a tested model exists which can be applied and replicated in an effort at planned development.[30] The 'green house' approach is based on the assumption of the 'slight stimuli' and 'Incentive' would be capable enough to generate an organization which will accomplish tasks of common interest.[31] The obvious limitations of first approach is that it is mechanical governmental, blind to peoples aspiration and interest of the people. On the other side, the green house approach is based on local interest, aspiration and capacity of totally based on the political economy of local area. It has a potential of strengthening common interest, mobilizing the hidden resources and developing agencies to chanelize even government efforts of development. Though the Indian experience of Replication of Anand pattern is not totally fitting to 'Blue-print' approach, due to number of differences. It has remained quite controversial issue in India.[32] It is quite interesting to note that recent study of Tushaar Shah has made some remarkable points in defence of the blue-print approach. He believeds that unless external catalysts, agent is available, it is unlikely that suitable collective actions may come about easily on its own. (Moreover, such approach will prove fast exceedingly 'slow' process for institutional development against their blue print approach when it works induces rapid development of local organizations.[33]) More interestingly, Tushaar Shah explains the true significance of "Foundation design" in an analogy of "Foundation Seeds", In his words, "We would like to argue that an organization can be designed for higher performance of certain kind just as plant breeder can breed for higher fertilizers response of high drought response. Moreover, just as foundation seeds can be used to multiply plants with certain configuration, so also can careful understanding of a foundation design can be traced to multiply co-operation or any local organization.[34]" However, Tushaar Shah has clearly prescribed the basic condition of the successful outcome of blue-print approach namely (a) Quality of original design concept (b) Quality of replication process and (c) Adequacy of minor adoptions. The limitation of the blue print approach is best explained in a statement. "if an external agency were artificially prop-up cloth merchant as a seller of utensils or a barber cum grocer, then use this a blue print for replication, the enterprise is sure to fail."[35] Similarly, circumstances of emergency of local organization, nature of business and social context of the local within which they operate are the factors which are required to be considered in adoption of blue-print approach.

Though, Tushaar's arguments by and large, are convincing but there is a catch for example, when we accept the difference of socio-economic, business environment to what extent one could become successful to adopt the design concept without the loss of the basic value of the approach is very difficult to understand. Moreover how could we judge the quality of design and replication process is also a pertinent question, empirically, non-adoption of Anand pattern even in some part of Gujarat and less successful replication of the same in other commodity like oil-seeds pinpoints the difficulties involved into totally depending on the "Blue-print" approach. Thus, both the approach have their limitations and it is difficult to arrive at final conclusions.

5. Who gets the benefits of co-operativization ?

Ideologically, co-operatives are supposed to help poor to Improve production system and co-operative should help not only production but also distribution and 'equilizing' the structure.) These questions of role of co-operation in inequality reducing process and development of Rural poor have been most recent issue in co-operative studies particularly after a remarkable success of co-operatives in the area of Sugar and Milk in Maharashtra and Gujarat. There are two extreme views ;one is that in a given socio-economic structure, no amount of tinkering with institutional frame work of the co-operative is likely to alter their basic condition. Moreover, it is believed that 'inequality' atleast moderate inequality, is not necessarily an obstacle to co-operative. Infact now a unique stand taken by some experts is that the objective of equality is not only unequal but also an imposed normative value from external sources over these concerning with co-operative.[36] However, there are adequate warnings against such arguments, will not such development lead to co-operative capitalism ? Will not such development destroy a protection of poor against the forceful marketization tendencies ? Particularly in the new environment of growing liberalization in co-operatives and declining state control which would lead to absence of protection of those poors who really need to be covered through co-operatives, until now, in the area of sugar and milk, there are ample evidence to show that small and marginal producers have got benefits of co-operatives, though, the significant proportion of landless poor are yet to be covered.[37] However, it is a fundamental issue of Indian co-operativization as to how to promote participation of Rural poor in the process of co-operativization ?

6. **What should be the Nature and extent of the role of the state in co-operativization ?**

One of the most debated question today is about nature and extent of the role of the state in co-operativization. It is said that the growth of the co-operative development,[38] In India is characterized by "Statism and infact, the current status of co-operatives in India is attributed to the so called negative role played by the state.

Historically, co-operative movement was initiated by the government as a relief programme . After the independence Gorwala committee recommended the greater role of the state by participation through investment in this sector which led growth of state finance to co-operative alongwith stable control. The limitations of the entry of the state in this way was well recognised long back, for example, Nehru made a bold statement recognising the dangers of excessive participation of the state even in 1958.[39] It is true that on account of the excess power to Registrar under co-operative. Acts, instead of state becoming "facilitor", has become controller and Manager. The state can also play a positive role to protect the interest of weaker sectors, protect the co-operative from vested interest and look after the state research invested in co-operative. (However, the present approach and extent of involvement and acts of co-operative laws obviously do not allow for greater initiative of the people. Member's participation and control on the cooperative, State has thus, become responsible for the exceeding politisation in co-operatives as state leaders are capable of influencing state participation in co-operative, according to their political needs). Thus, ultimately, instead of members becoming their political leaders who also called themselves co-operative leaders have in the words of Kurian, hijacked the whole movement for their narrow and nasty benefits.[40] In fact, the current co-operative leadership which is not only corrupt but also characterized by contrary values to co-operative principles is the most responsible for creating a public image about co-operative which is not very encouraging, despite the great potential it carries to promote people's welfare.

7. **What Issues will be required to revolve due to co-operative liberalization ?**

Recent New economic policy of adjustment and restructure, has affected co-operative sector also. As a part of the opening up the liberalization has been introduced in co-operative sector first step in that direction was deregulation of dairy industry. We have discussed at length the logic and consequences of the delicencing dairy industry elsewhere.[41] However, this current new economic policy will lead to liberalization and restructuring of co-operative sector. The obvious implications of this policy will mean (a) competition (b) withdrawal of reservation and protection (c) withdrawal of state finance and investment (d) withdrawal of state from direct sponsorship and support preferences etc. These steps designed under the new policy challenges not only to growth but even very existence of the co-operative sector even in those areas of economic activities where co-operatives have shown remarkable record of performance. Unless, very specific actions are being undertaken by the state to restructure co-operatives adversity arising out of present economic policy,(defeated the very objectives of new economic policy will get)

Thus, these 'Issues' are perhaps the basic factors responsible for crisis of co-operative movement in India.

Positive Approach, Initiative and Actions

The critical review of progress and issues make the task clear to draw a suitable policy package with an objective to revitalize and restructure co-operatives to give rise the robust national Co-operative Sector which would be very active in development, planning and economic functioning of the nation. Major steps to be taken are suggested as follow with the hope that this would help developing a typical co-operative sector which is most needed today in India.

1. National Co-operative Policy

In the line of Draft National Agricultural Policy, this sector needs a clear cut well designed National policy which would be a guide for the systematic development of co-operatives. This statement should spell out National Objectives, approach model of co-operative structure, major instruments and specific task of the states in development of co-operatives. This will resolve our many difficulties. The state government then may ask to develop their strategy based on the basis of such a clearly worded National Policy document.

2. National Initiative For Co-operative Development

National initiative for the development of public opinion, create awareness of real problems and solutions of current crisis of co-operatives and to develop a strong public pressure lobby for the vibrant co-operatives is highly wanted today. A powerful Public movement based on some basic lines of changes in its structure and approaches to initiate the process of revival and revitalize the co-operatives is essential. One such excellent example is a great partnership of voluntary agency and government organization under the guidance of respected co-operative leaders.[43] This co-operatives initiative project showed by 'NDDB' and 'Samakuya' and NGO has provided a great guidance and opportunity for all those who wish to have meaningful participation in reforming co-operative sector. They have an approach to develop a committed constituency of co-operative practitioners across the country and those have knowledge and faith in co-operatives and a panel of open leaders and their linkages to help promotion of dialogue on the issue faced by co-operatives. The actions involve identification of 1000 practitioners of co-operatives, series of background papers, organization of National, regional and local workshop to maintain 'dialogue' and develop a 'New Hope' for co-operatives as "people's vehicles" of development. Nationally respected leaders and devoted co-operatives practitioners like V. Kurien, L.C. Jain and Mohan Dharia who established National Panel has provided a great hope of revival, revitalization and restructuring of co-operatives which would sustain successfully its burden of task easily. The project within its three years of operation is most likely to be succeeded in strengthening co-operative movement and has provided an example of constructive approach for those individuals and NGOs who are concerned with the development of rural people through co-operative approach.

3. Promotion of Member's Participation

The greatest strength of any successful co-operative next to people's initiative is member's participation in co-operative 'control' is the key to promote "member first" co-operatives. However, merely an 'opportunity' by the law would not help. They are required to be prepared or rather adequately 'educated' to enable themselves to undertake such tasks in positive and constructive ways. NDDB once again provided a fresh, initiative and alternative of "co-operative development program" primarily involving people. It has been hypothesized by observing behaviour of 65000 dairy co-operatives and 450 oilseeds co-operatives that the successful co-operatives depend upon the member's awareness, their participation democratic governing and proper institutional management.[44] This is impossible to achieve unless we have very dynamic program for 'members education' which would lead to change in the attitude, behaviour and responsibility unlike uncoordinated scattered governmental programme, NDDB has devoted a programme focusing on all 4 segments of co-operative namely members, management, professional and social political environment. The specific tools developed are relating to members education programme, women's education programme, managing committee training and seminar, secretary orientation Programmer, youth farmers and women clubs and leadership Development Programmes.[45] This kind of programme of NDDB initiative by NDDB is a pilot project which will extend for 5 years. In a short span of three years, the project has shown 'some positive' results.[46]

4. Abolition of antiquated co-operative societies Acts and Adoption of the Model Act of Co-operative.

Co-operative needs bold steps to reform and restructure. One such step is immediately to abolish an antiquated, colonial co-operative Act which has been instrumental to convert 'Sahakari' Movement into "SARKARI" Movement and making 'Registrar of co-operatives' 'Brahma, Vishnu and Mahesh' The very lucid and detailed account of co-operative legislation and its consequence have been well stated in recent study of S.N. Singh.[47] It is sufficient to note that instead of expressing the 'Principles of Co-operatives' it became instrumental to kill all spirit and values behind these principle and infact, instead of becoming promoter and protector of the co-operative. It converted itself into controller and manager. Further, such co-operative laws are absolutely outdated in the context of the co-operative development recorded in the last two decades ? Recently, drafted Model Act of co-operatives provides a real alternative and guidance.[48] Every effort should be made to see that all states enact the act on these lines and patterns of Model. The Act is expected to reform and restructure the co-operative due to its distinguished features. For example, it spells out specific role of the state suggests simplified procedure for registration of co-operatives, takes away all anti-democratic powers from government and also restricts all powers of registrars, such as super session of the Board of directors, compulsory amalgamation or division of societies, compulsory amendments of the buy Laws, veto/rescind/ annual resolution. It is also notable that as a step towards democratic governing it provides many powers to federation of co-operative. They will be responsible to see that elections are held and audits are done in right time by the co-operative societies. Thus, 'Model co-operative Act' has a great potential to revolutionize the structure of co-operatives system in India and hence, people's movement and political lobbies must be developed to pressurize the state to pass such Acts at the earliest.

5. Development of Federation/Unions

The model act provides great role and power to federation/ union. Moreover experience of dairy co-operatives development in Gujarat and number of states like Tamil Nadu, Andhra Pradesh have shown that they can play a crucial role in development of co-operatives however, do we have adequately developed federations and unions in states for all types of co-operatives ? To carry out the changes in the right form, therefore, development of Federation and unions on the line of GCMMF and 'Kheda' union are highly required, otherwise, the envisaged changes will be just on "Papers" and In 'politics' but not in the process of people's control on institutions.

6. **Constitution of co-operative Tribunals for Corrupt practice co-operative crimes and adjudication of disputes including appeals under the Act.**

The most serious are the co-operative crimes in the form of misusing funds and corrupt practices which have tarnished the image of the co-operative among public. The regular legal procedures are lengthy, complex and time spending, hence, to correct this source of weakness, an idea of constitution of co-operative Tribunals originally provisioned in the Model Act for the adjudication of disputes must be extended further. Such co-operative tribunals should be available to every district with the judicial officer in the rank of a district judge.

7. Strict Enforcement of Milk and Milk Product Order - 1992

It was in 1991, when Government declared the revised list of reserved industries, Dairy product manufacturing sector was excluded and thus deregulated, we had made a detailed analysis of this measure and suggested some measures to save the co-operative dairy industry. One such measure was relating to some kind of Board to regulate the liquid milk supply to protect consumers on the lines of UK.[49] Ultimately, the grave error in unregulated opening of the doors of dairying sector was realized and on June 9, 1992, the order under the essential commodities act aiming to increase in supply of liquid milk to public and to prevent its diversion to make high value milk products. Though, private dairies may state it as "backdoor licensing move" the MMPO June, 1992, requires registration for all units handling Milk product, containing milk solids in excess of 500 tonnes per annum. The provision of milk and Milk product Advisory Board with controlling powers over registered units are adequate enough to assure liquid milk supply to consumers and also to protect co-operatives from undue competitive forces within their milksheds. The MMPO must be followed completely and without any dilution. This provision is one which has regulated the cautionless co-operative liberalization.

8. **New provisions to encourage co-operatives for Maximizing benefits of Liberalization policies of Trade financial and International trade.**

The new economic policy of restructuring and stabilization has certainly open up opportunities of development on the part of those non credit co-operatives which are very large units comparable with any large private or multinationals. GSMMF, KRIBHCO, etc. illustrate this kind of co-operatives, nor can they protect them from competition from national and multinationals, unless they also are give certain power by relaxing existing provisions. For examples, when state intends to withdraw their funds from co-operative why the co-operative should not be allowed to register their subsidiaries to mobilize funds ? Why the co-operative company should not be allowed to formulate ? Why should we not make required tax reforms to enable co-operative to spare their strength after payment of taxes ? Why should we not make essential changes in the Labour Laws to enable the co-operative to compete with private sector.[50] The point is that, the changes in the business environment has compelled to change the co-operatives also but the changes required by co-operative can not take place as the power for the changes are lying with government, particularly due to old co-operative Acts ? Thus, unless the new provisions are made to enable to co-operative will not only be unable to grow but even found difficult to sustain itself. Therefore, to do justice, atleast, co-operative should be empowered with the liberal provision in the areas above stated.

9. Development of co-operative professionalism.

The Natural and size of co-operative development in India denies now simple political ordering and trading methods of managing co-operative sector now equally needs professionally committed managers. They are supported to be committed to co-operatives and sound in professional management with due practical understanding of the sector. Also what is required now is to develop a management culture. The organizational system will be required to be developed their the members, Board of directors and managers all will have a very demarcated functions to play without encroaching upon each one's power with clear cut responsibility of the actions. This is the lesson provided by the successful co-operative enterprises.[51]

10. Actions to Promote Viability :

The unreliable and inefficient co-operative can not provide an alternative to private trade or it could secure the member's patronage. The economic tangles leading to unviability such as higher unit cost, higher service charges, losses, overdues, linkages via corrupt practices; low prices of products, narrow margin of trade, deliberately drawn lower price spread etc. will have to reasonably resolved while the measures suggested by recent Khusrow Committee are worthy implementing; Bold initiative and efforts will be required to introduce "restructuring process" with a variety of packages of actions suitable to different situations this maybe included tied contracts,

Amalgamation, Reorganization of units; quality control measures, new market opportunity, drives, Diversification, rotation system of internal control and incentives, amendments in the relevant economic legislation etc.[52] However, a point of caution is to be noted that there could not be one single strategy but many strategies which would be designed on the basis of analysis of the specific circumstances faced by different size and type of the co-operatives.

11. Participation of Poor in Co-operatives.

It is true that one may expect a miracle to happen with regard to fundamental characteristic's of economy; the deliberate efforts are needed to involve poor producers in the compass of co-operative. The three tier Anand pattern model has its great utility in this regard. There are large number of evidence which have shown how Rural poor benefited from Anand pattern dairy co-operative.[53] Recent study have shown that replication of Anand patten co-operative is very encouraging[54], however, many more design concepts are needed to cover variety of people and economic activities, may be based on the same Anand pattern co-operative work with great modifications, however, the question of developing co-operative for poor is not merely finding out a suitable model but also involves number of issues, experts like Uma Lata[55] and Anil Gupta[56] who have done substantial works on these issue. There are atleast two major messages. (1) If we want develop co-operative for poor, we should have adequate knowledge of actual constraints and steps needed to alleviate this constraints. Our failure to develop co-operatives for poor indicates that we donot have the adequate knowledge of the constraints and solutions. (2) Another point is of worth nothing, where Anil Gupta has reminded us of our poor knowledge and our unwillingness of learning from Traditional Co-operatives which would help us to develop co-operatives for poor in relatively easy and acceptable ways than what we are trying today with modern co-operative models. It seems that, this is also an approach worth consideration when we attempt to promote participation of poor in co-operatives.

CONCLUDING OBSERVATION

Thus, the critical review of the Co-operativization in India despite the discouraging attitude and actions of Government do provide a hopeful scenario which could be appropriately develop with the line of actions and Approach suggested in the paper, however, we do know that this is how we can overcome the crisis but do we have "DETERMINATION AND DARING POWER" to do it ? The ultimate design of the co-operatives institution will very largely depend upon the kind of response to this question will be made available by all of us in future.

REFERENCES

1) **Government of India :**
Planning Commission : - First five year plan 1951.

2) **D.R. Gadgil** :- Towards a co-operative common wealth Prof. Brij Narayan Memorial Lecture, 1960 published in Speeches of Prof. D.R. Gadgil on 'co-operation, Gokhale Institute of Politics and Economics, Poona, 1975.

3) **Indian Co-operative Alliance** :- Annual Report - 1985.

4) NABARD :
Important Items of Data :- Credit Non-credit Co-operative Societies, 1990-91.

5) **J.S. Crag :**
Indian Co-operative in the 21st Century, Breaking Bureaucratic Noose ! International Symposium on Management of Rural Co-operative, a distinguished lecture, 1992, IRMA.

6) N.S.S. :
Debt and Investment Survey of Rural and Urban Households, 1992.

7) **B.K. Ghosh & K.V.Patel :**
Institutional Credit for Farm Sector, marginal and Small Farmers IASSI quarterly, Vol. 9, No.4, 1991.

8) Ibid Ghosh and Patel pp. 69.

9) **S.S. Puri :**
Critical issues for the co-operative sector, IASSI quarterly, special No. on the co-operative sector Vol. 9, April-June, 1991, No. 4 pp. 20.

10) **Ibid S.S. Puri 21.22.**

11) **Dilip Shah :**

(i) Co-operative Dairy Farming and instrument of Rural Development, 18 Jan. 1989.

(ii) Anand Pattern Rural Development, Kurukshetra Feb. 1986.

(iii) Dairying : An Instrument of Rural Development, Journal of South Gujarat University, Vol. XI, 1982, pp. 114-116.

12) V.M. Dandenkar :
Co-operatives and Rural Development - Keynote Address to the Third National Convention on Studies in Co-operation at Waltair, 1982.

13) B. Vankatappaih :
Plea for Studies in Ideologies and Practical Realities in co-operative Development. Keynote Address to the First National Convention

on Studies in co-operation, June 1980.

14) Ibid Vankatppaih pp. 9.

15) Tushaar and Others :
Design of Democracy, Building Energetic Farmers Organization: Workshop Paper (7-13), Symposium on Management of Rural Cooperatives (Draft) - 1991, pp. 4,5,6.

16) Tushar Shah :
Management of Village Dairy Co-operatives - Field Notes from Laxmipura (Sabarkantha Dist.) IRMA.

17) Tushar Shah Ibid Design for Democracy pp. 6.

18) B.S. Baviskar and D.W. Atwood :
Fertile Grounds : Why co-operatives Flourish in Western India ? IASSI Quarterly Vol. 9, No.4, 1991, pp. 82,83,84.

19) Tushaar Shah
Learning from Success : Analytical Review of Case Studies of Successful Farmers and Farmers Co-operatives - Rapportiers Report, 50th Annual Conference of Indian Society of Agricultural Economics, 1990.

20) W. Mohanan, C. Krishnagopal, S. Mahalingan :
Development of Monitoring and Evaluation Mechanism for Impact of Co-operatives - A study sponsored by CIRDAP, NDRS, Mimograh - 1991 (Unpublished).

21) Baviskar - Atwood - Ibid pp. 82.

22) Baviskar - Atwood - Ibid pp. 84.

23) Tushaar Shah :
Design for Democracy - Draft Paper circulated IRMA Symposium 1991.

24) Tushaar Shah :
The Logic of co-operative company, IRMA, 92.

25) Ibid Tushaar Shah : Design for Democracy.

26) Deviprased Misra and Tushaar Shah :
Analysing under performance in village co-operatives, working paper : (22) IRMA, 1991.

27) Ibid Misra-Shah pp. 4,5,6.

28) See for example :

(i) Shanti - George : India's Operation Flood - Oxford 1989.

(ii) Dilip Shah : Replication of Anand Pattern Dairy Co-operatives, **Financial Express** Feb. 1990.

(iii) Martin Dornboss and others : Dairy Aid Development, Sage, Vol. 1, 190.

(iv) Dilip Shah : Replicability, Dependency and Equity Issues in Operation Flood Project, **Indian Journal of Political Economy.**

29) Godan Hyden :

Approach to Co-operative Development : Blueprint Vs. Green House in Atwood-Baviskar (edited) who shares co-operatives and Rural Development, Oxford, 1988.

30) Ibid pp. 155.

31) Ibid pp. 158.

32) See for example :

(i) **Shanti - George** : Co-operatives and Indian Dairy Policy More Anand than pattern in Atwood-Baviskar Ibid.

33) Tushaar Shah :
Energizing India's Rural Co-operatives - Draft Paper circulated before IRMA Symposium, 1991.

34) Tushaar Shah Ibid, pp. 8.

35) See for example :

(i) Atwood - Baviskar : Who Shares ? Co-operative and Rural Development - Introduction pp.5.

(ii) Tushaar Shah : Ibid Design for Democracy, IRMA, 1991.

36) (i) K.T. Acharya, Vinod Huria : Rural Poverty & Operation Flood
(ii) Dilip Shah : Dairy Farming and disadvantaged groups, **Economic Times**, January 1992.

37) S.J. Phansalkar and Akhileshwar Pathak :
State Co-operatives and Inter-relationship - A Term Paper in Workshop on State and Rural Co-operatives, IRMA, 1992.

38) Nehru Jawaharlal :
Third Indian Co-operative Congress Inaugural Address of Pandit Nehru, 1958.

39) V. Kurien :
Co-operative Leadership and Co-operatives values : Co-operative Perspective, 1991.

40) Dilip Shah :
Delicensing Dairy manufacturing Sector : Logic and Reality, **Financial Express**, 31st May, 1992 (Two parts).

41) See number papers :

(i) Taimini K.K. : Co-operatives and Structural Adjustments, Economic Times, 26th Aug. 1991.

(ii) P.R. Dubhasi : Opening up the Co-operatives, Financial Express, 3rd June 1992.

(iii) P.R. Dubhasi : Co-operatives and New Economic Policy, Financial Express, 18 Nov. 1992.

42) Co-operative Development Group, NDDB :
Co-operative Initiative Project, Press Release, August 22, 1992.

43) Arun Wayangankar and S.N. Singh :

Co-operative Development Programme of NDDB. Paper presented in IRMA Symposium, Workshop on Co-operative Education, Dec. 1992.

44) Ibid.

45) Ibid.

46) S.N. Singh and K.V. Ramaseshu :
Trends in Co-operative legislation in India, co-operative Development Group NDDB, 1993 (Unpublished Paper-Draft).

47) Planning Commission, Government of India :
Report of the Committee on 'Model' Co-operative Act, May, 1991.

48) Dilip Shah Ibid Delicensing Dairy Product Manufacturing, **Financial Express**, Part-II, 18th May 1993.

59) V.Kurien :
Five decisive steps - Note to Prime Minister, Aug. 1992.

50) Tushar Shah Ibid Learning from the success : Analytical review of case studies.

51) Tushar Shah : Ibid.

54) G. Parthsarthy :
White Revolution, Dairy Co-operatives and Weaker sections EPW 28th Dec. 1991.

- Acharya K. T. and Huria V. : Rural Poverty and Operation Flood EPW, Sept. 1986

- Mohanan : Impact of dairy Co-operatives on Poverty groups in the Operation Flood, NIRD, 1986.

53) Dilip Shah : Ibid
Financial Express, 2-3 Feb., 1990.

54) Uma Lela :
Co-operatives and the Poor : A Comparative Perspective Development, Vol. 9, (1981) pp. 55-72.

55) **Anil Gupta** :
Why Poor People Don't Co-operate : Learning from traditional systems in G. Glare Wanger (edited) The Research Relationship, 1987.

□□□

2

Role and Future of the Cooperative Movement in An Environment of Economic Liberation

INTRODUCTION

Outlining the various reforms undertaken by the Government of India is leading to the new economic environment for the cooperatives this paper explains that some policies may result in dislocation of economic resources and priorities. To counter the same, 'safety-nets' have been suggested in the form of anti-poverty programmes, direct employment generation programmes etc. As the co-operative institutions in India are not homogenous, the application of the new economic policy will not only have to be differential, but also disproportional.

Co-operative Development Scenario

A quick but critical review of the co-operative development in India brings to notice the following underlined features of co-operative development in India.

1) India's Co-operative Sector is the second largest sector in the world next only to China in terms of membership. The size of the Co-operative Sector can be judged from the fact that there is an unignorable number of 3-5 lakh co-operatives spreaded in villages with coverage of 145 million people and a working capital of Rs. 48,000 crores.
2) India's Co-operative Sector has transformed into a strong diversified sector. For atleast 60 years, it was just a credit dominating co-operative activity limited to the agricultural sector. This has substantially changed to the coverage of a wide range of diversified non-agricultural activities. Statistics pertaining to the year 1989-90 shows that 60% of the co-operatives are in the non-credit sector with a diversification into 38 types of non-agricultural economic activities.

The quantitative achievement in terms of hard facts state :

a. More than 50% institutional credit is distributed by the co-operative sector.
b. Around 60% of main input responsible for the green revolution, namely fertilisers is distributed through co-operatives.

c. From a mere Rs. 13 crores worth of agricultural produce, co-operatives now market Rs. 5530 crores of the same.
d. Almost 60% of the sugar in India is produced by co-operatives.
e. Market share of milk distributed in many cities is 90% to 95% with a national coverage of 21% of production by co-operatives.
f. Co-operatives account for 21% of wheat, 18.5% cotton and 32% of the jute procurement.
g. Around 26,000 housing co-operatives completed 7 lakh houses for the poor and middle classes.

3) The Indian Co-operative Sector has also shown a kind of 'dualistic' performance scenario wherein some co-operatives confined to specific commodities like that of milk, milk products, sugar, fertilisers, etc., have shown great viability, vitality and a significant market share which is in no small measure comparable to cooperate business enterprises. However, there are large number of co-operatives whose record of performance is very discouraging.
4) There are 5 States namely i.e. Gujarat, Maharashtra, Tamilnadu, U.P. & M.P., where 80% of the co-operatives are concentrated. There, however, exists a remarkable variation in their performance in terms of region, commodity and variety of activities.
5) Finally, despite the fact that co-operatives have achieved their objectives of rendering service to the weaker sections of the society in areas of credit, fertilisers, milk, sugar etc., in some States, the public image continues to be that of an official or a "SARKARI" movement. It has still to become a people's movement in the true sense of the term. However, despite such limitations, the performance of the co-operative movement is not so discouraging so as to not even deserve a conventional place in the Eighth Plan document of the nation. Infact, it is widely believed that "the adoption of the new economic policy has not only marginalised the importance, but has virtually made the co-operative movement irrelevant." Instead, the importance of co-operatives as the only alternative institutional safety net for the weaker sections against the forces of non-friendly market orientation, privatisation and liberalisation is emphasised under the regime of the new economic policy.

Economic Liberalisation and the Cooperative Sector

The economic liberalisation adopted by India since June, 1991 has two phases of policy actions viz. the policies concerning stabilisation and those related to restructuring of the economy. The important characteristics of the new policy relate to :

a. Minimum Government control and interference by way of withdrawal of the State.
b. Deregulation of areas of production subsidised and reserved for the co-operative sector.
c. New trade policy oriented to internal and external competition.
d. New Industrial Policy.
e. Exit policy for restructuring of the employment system.
f. Reforms in the financial sector.
g. New international trade policy encouraging exports and imports.

These policies would result in a dislocation of economic resources and priorities. As a result, to minimize these effects, "safety nets" are suggested in the form of anti-poverty programmes, direct employment generation programmes etc.

Impact of the Policies

The impact of these policies on the co-operative movement depends on the nature of the co-operative movement itself. Cooperative institutions in India are not 'homogenous'. They are characterized by a number of differences. Therefore, positive or negative impact of the new economic policy will not only be differential but also improportional. For example,

a. The impact of new economic policy on strong, self-reliant autonomous co-operatives will be negative but development could be ensured by deliberate policy action which provides the benefits of the new economic policy to co-operatives. However, impact on state dependent co-operatives will be very adverse.
b. The impact of new economic policy will be adverse on co-operatives which have just been initiated in relatively weaker co-operative areas.
c. The impact of the new economic policy will also depend upon India's approach related to the adoption of the capitalist system, socialistic system or to adopt a "middle path" in their policies. Prime Minister's a recent ddress at DAVOC has highlighted the fact that India would stick to the Nehruvian policy and adopt a "Middle way" rather than limitless capitalism. If that is so, co-operatives will have a much greater role to play as truly envisaged in the Nehruvian Model of mixed economy as the third sector of the economy. However, the actual economic policy followed until now does not support the possibility of a "middle path" policy.
d. Finally, the short run consequences and long run consequences of economic reforms would not be the same. If appropriate policy

actions are taken, long run policy impact could be positive to the co-operative movement in India.

Areas of Development Opportunities to Cooperatives

Development opportunities in the present economic liberal policy could be pinpointed in terms of the following :

a. The efficient and large co-operative units will have better scope of growth due to competition giving rise to need for constant efforts of locating new markets, products and technology for not only existence, but also 'efficiency' in terms of better quality products in adequate quantum at reasonable price. The rate of returns for member farmers will have to be changed from raising the sales price to reducing the cost of production and consequently raising the returns to producers.

b. The co-operative marketing organisations will have greater role as they have a much wider presence as compared to the corporate marketing companies especially in remote rural areas.

c. With the introduction of financial reforms and a consequent cut in priority finance, the role of credit co-operatives will be further enhanced. Infact, with increased opportunity of investment, co-operative financial institutions will have a role to play in the area of non-advance activity.

d. With a liberal policy for imports and exports, the scope for exports of agricultural co-operatives in some crops, vegetables, fruits etc. will be enhanced. The GATT agreements, if really followed by a cut in the 'subsidies' particularly in the area of milk and milk products, there will be an excellent opportunity to market at competitive price, not only in the developed countries, but also in the developing countries.

e. The tremendous opening up in the agro-based processing industries at large will also provide co-operatives an added advantage, being indigenous and strongly rooted organisations in rural areas. With merely 3% output of fruits processed in India and merely 20% milk processed by organised dairies, despite the entry of the private sector and MNC's there will be adequate scope for the co-operatives also.

f. There are a number of areas where the role of co-operative organisations as an instrument to compensate for negative side effects of structural and adjustment policies will be emphasised. These areas are, food security management by co-operative PDS, maintenance of rural infrastructure, schools, hospitals and other social service organisations which will be less attended in the present environment of deregulation and withdrawal of State.

g. Finally, co-operatives will provide a cover to most weaker sections in number of areas such as consumer's stores, credit co-operatives and even management of anti-poverty programmes to optimise the benefits of income and employment for the poor and weaker sections.

Economic liberalisation if applied to co-operatives in the true sense with gradual and strategic phasing of changes, it could provide a boost atleast to those which have proved a great success in terms of their performance. At the same time, we have to think of those who are not autonomous, but State aided units which are required to survive for the reason of protecting weaker sections. A separate policy will be needed to safeguard problems arising out of present economic environment and economic policy when applied to them. In view of these situations, it is suggested that the following main policy agenda to serve the needs of all these three different kinds of co-operatives prevailing in India will have to be initiated.

Steps concerning Corporate styled Autonomous Cooperative Enterprises

Those co-operative enterprises which have become successful are not due to regional culture, sound leadership and management or absence of bureaucratic and political interference but due to the fact that they are successful in meeting the interest of the members following the 'members first model' rather than just doing lip service to the 'value first model' dealing with equity and justice etc. Such co-operatives could do better and even take advantage of competition with the private sector and even MNCs, provided, there is a "level playing field" in terms of competition, by the way of some immediate policy support. It is obviously 'unfair' to assume that co-operatives have to follow all rules of the game while the corporate sector can afford to ignore all rules within a legal structure. What is urgently required is to formulate a "National Cooperative Policy" on the lines of the new economic policy extending economic liberalisation to co-operatives also.

a) Colonial Cooperative Legislations must be replaced by the Model Act

The co-operative movement in India was a part of the administration of the colonial Government, like our present 'MECHOLE' education. The co-operative laws in India are still unchanged. The Registrar of Cooperatives in India is the Bramha, Vishnu and Mahesh of the co-operative movement. It is rightly viewed that 'any co-operative law in any State must be in accordance with the provisions of Articles 19(4) and 19(6) and any provision contrary to this is violation of fundamental rights as guaranteed under the Article 19(1) (C) and (D) of the Constitution of India. This has hampered the process of an autonomous, self-reliant co-operative movement and leads to 'SARKARI COOPERATIVES'. Under the new economic policy, the co-operatives cannot compete or even survive if the Cooperatives Act is not changed. The recent ILO studies devoted to determine the framework within which self-reliant co-operatives can develop have concluded that 'in most cases present co-operative legislation is deficient'. It also notes that some African countries (Cameroon, Tanzania and Uganda etc.) which are fairly advanced, have already changed their co-operative laws and some other countries like Indonesia, Malaysia and Philippines have also changed their laws, reducing State control and transferring the promotional functions to co-operatives.[3]

In India, the Chaudhary Brahmaprakash Committee appointed by the Planning Commission drafted the Model Cooperative Societies Act which has been approved by NDC and also incorporated in national Cooperative Policy, is in consonance with the fundamental rights of citizens and present needs of the time. The State Cooperative Acts must be accordingly changed suitably in India at the very earliest.

Further, in view of the conflicting interests, even among co-operatives of States, it is time that the Government of India enacts a new 'Multi-State Act'.

b) Cooperative Company must be formulated

Another step in the direction of balancing the strength of co-operatives with companies is to make an amendment towards establishing co-operative companies as per the Companies Act. The idea of a co-operative company is essentially an ideal response to help co-operatives of the size of AMUL to compete with MNCs.

The co-operative company like any other company will be member organization. However, it will be governed and managed primarily with a view to practice and maximise the patronage interest of the members. This will help co-operatives to execute capital by equity shares, facilitate greater returns to investment and thus help promotion of further investment. However, it is believed that the concept of companies is contrary to co-operative ideology. This fear is not well founded. Acceptance and effective implementation of the fact that "the co-operatives are in the first place private enterprises in the business sense", other principles of co-operatives will follow. Immediate. In developed countries like USA, UK, New Zealand, Zimbabwe, Canada etc. it is a normal practice that business organisations which operate on the co-operative principles are often incorporated under their relevant companies Acts. Even in our country, there is already a well established example of a farmer owned company which has taken the first step in this direction. The Agro Complex India Ltd. registered in 1983 as a public limited company is owned by around 7000 poultry farmers. Anyone who has atleast 250 birds can become a member by buying a share of Rs. 10. This ensures that non-users cannot become members and thus the fear of absentee control is taken care of. This poultry farmers company now has a share capital of around Rs. 50 lakhs, projected annual turnover in 1992-93 of Rs. 45 crores and a daily sale of Rs. 15 lakh eggs. It is suggested that a formal provision needs to be made in the Company Act which allows the co-operative company to establish some kind of a loose proportional linkage between allocation of the voting rights with the number of shares owned and with the amount of business done with the co-operative.

c) Change in the Tax Laws in favour of Cooperatives

A recent national convention has deliberated on this issue. A review of member-co-operative transactions has to be made for tax purposes. Tax laws must appreciate that where a co-operative deals with non-members, it may be treated as other business areas. However, where a co-operative deals with its members, the transaction is not of purchase but of pooling (as in producer co-operatives) and not one of sale but of distribution (as in supply co-operatives). Co-operatives should be required to maintain members and non-members accounts clearly, and tax laws should cover non-members transactions only. It was noted that since co-operatives too, have surplus and distribute this to members, member's individual income may be taxed as in some western countries. Simple changes in the tax laws will ensure a better capital base and relieve the Government to provide grants and loans out of the tax returns collected from co-operatives themselves.

d) Modification of Labour Laws applicable to Co-operatives

The present laws are not responsive to the needs of producers. Sometimes, it becomes a great burden and an unjust burden to respect the provision which also tightens the hands of the management in operating the 'exist' and 'hire and fire policy'. The labour legislation should be modified to ensure that meeting the needs of employees should not be at the expense of the rural producers.

e) Development of Federation/Unions

The strong, competitive, vibrant co-operative structure which can face competition, calls for the development of a three tier structure in the true sense. In the areas where co-operatives have developed very well, the Federation as well as the Unions are not strong. A kind of bargaining power in the market could be acquired by a strong federation which also regulates competition among co-operatives and coordinates the system against private competition. The Model Act provides a greater role and power to the federations. With a gradual withdrawal of the State, strong federal structure will be required to develop viable co-operatives as also a voice for the whole industry.

f) Resource Mobilisation Drive

With the withdrawal of the State, grants and aids resources will be grave problem. In this context, some important recommendations of the high power committee by Maharashtra has made very valuable suggestions to raise share capital from the general public. In this, persons who subscribe to preferential capital would be treated as non-members without any voting rights. However, the committee feels that it maybe possible for the co-operative sector to attract public financial support by way of share capital if the rate of returns on capital is sufficiently attractive.

Another way of resource mobilisation for co-operatives, is to raise term deposits from the public. A co-operative society even under the existing law can accept deposits from persons who are not its members. There is no legal restriction in the co-operative law on acceptance of deposits from the public. Similarly, debentures also could be floated by co-operatives. Thus, some known conventional schemes of resource mobilisation could be worked out.

g) Social control on Co-operative & Corporate Sector in the socially significant areas of production

Social regulation is an essential instrument to protect the social interest in some sensitive areas of consumption and production. Dairy, sugar oil etc. are such areas where unregulated liberalisation could be very damaging and hence the policy instrument devices such as Milk and Product order 1992 are steps in the right direction, whereby not only co-operatives, but also corporate units are adequately controlled and regulated.

h) Qualitative control mechanism for exportables

With opening of the economy, it is possible for the export potential to be realized. There is great scope for exports in the areas of processed fruits, vegetables and dairy products, provided we have exportable surplus. However, even after acquiring adequate surplus products, we must have a regulative and canalizing central mechanism to control quality and maintain standards which are not easily retained. The experience of NDDB in case of milk products could be utilised in case of other exportable commodities also.

i) Development of Professionalism in the Co-operative Management

The most delicate challenge faced by co-operative enterprises is to deal with dilemmas arising out of contrasting interests and values. How to reconcile these problems is a challenge before professionals. They have to maintain efficiency of enterprise without loss of ethics. They have to balance the producer's interest without unduly exploiting consumers of the society. Apart from this, with increased and open competition of non-co-operatives, a lot of professional capacity will be required to develop markets,interpret market signals, coordination among co-operatives and to utilise internal opportunities. These co-operative enterprises which have recorded success are already on the way to design and develop such professionalism.

j. Human Resources Development

A recent workshop on the issue of co-operatives and the new economic policy has well analysed the role of HRD. Changes in the working environment affects not only the managers, but all related to the development of co-operatives. Therefore, under the special efforts of training and education, the focus of the HRD strategy should be on members, elected leaders and employers, NDDB has done model work in the area of HRD in co-operatives. Their efforts should be extended to others and the process of HRD promotion should continue. Thus, the above stated package of actions, aims at enhancing the capacity of co-operative enterprises to compete with any non-co-operative organization without much difficulty ensuring its continuous and faster progress in India.

k) Steps concerning Co-operative Development for the poor in the poverty zones for the promotion of co-operatives in social areas including developmental programmes.

The correct scenario of co-operative development of India lies much in the success story of co-operative enterprises in the States like Gujarat, Maharashtra, Tamilnadu, A.P. and Karnataka where 80% co-operatives work in milk, sugar, cotton and credit. However, the co-operative movement meant for the rural poor must not only survive, but be strongly promoted and it also should provide adequate institutional cover of adjustment and restructuring. This task could be done by a changed role of the State as well as the attitude and strategy of the State in the promotion of co-operatives for the poor. (These efforts could not become successful unless, NGO's are encouraged to organise new co-operatives, revitalise old co-operatives and promote co-operatives as the people's movement by taking fresh initiative in co-operative development on the lines of NDDB approach where the actions involved identification of 1000 practitioners of co-operatives, series of background papers, organization of national, regional and local workshops to maintain a dialogue and generate a new hope for the development of real co-operatives serving the needs of members first. A National panel of respected and devoted leaders like Kurien, Dharia and Jain not only successfully highlight the needs of the new policy, but also help in generating a new confidence in co-operatives. Apart from developing co-operatives on participatory lines the public image of co-operatives needs to be strengthened. This task calls for specific actions. For example, it is suggested that co-operative tribunals should be established. This is due to the fact that the co-operative crimes in the form of misappropriation of funds and corrupt practices which have tarnished the image of co-operatives amongst the public are larger and occur at frequent intervals. As the regular legal procedures are lengthy and complex but also time consuming an idea of constituting tribunals originally provisioned in the Model Act for the adjudication of disputes must be extended further. Such co-operative tribunals would be available to every district with a judicial officer in the rank of a district judge.

Thus, what is required is to develop co-operatives which are :

* Self sponsored
* Accountable leaders/employees to producers/members
* Producer sensitive
* Autonomous
* Enlightened membership with patronage

Concluding observations

The critical assessment of the co-operative sector and new economic policy lead us to the following observations :

The co-operative sector in India is peculiarly developed where we find three types of co-operatives (a) autonomous (b) **State aided and dominated** (c) vested interest oriented. The role and relevance of such co-operatives will, therefore, not be the same. **The effect of new economic policies will also not be the same.**

Therefore, powerful cooperative structural calls for not total withdrawal of State but very strategic role of the State to help developing cooperatives of the poor and strictly controlling while completely withdrawing from autonomous self reliant and strong cooperative enterprises).

The present economic liberalisation policy calls for judicious application of liberalised policy to co-operatives and to provide a rightful strategic place as a countering mechanism to harmonise the hazards of privatisation. It must be noted that the new economic reforms will make no sense without development of relevant institutions and total transformation in the fundamentals of society.

References

1. Ministry of Finance, Two years of economic reforms 1993.
2. Prime Minister's Address of DAVOS. The Middle Way, Mainstream, February 1994.
3. B.M. Myth of a Middle Path, EPW, March 5, 1984.
4. L.C. is Narsimha Rao follows Nehru's Footsteps ? Mainstream Aug. 6, 1994.
5. V.S. Vyas, Agriculture policies for the Nineties, EPW, July 25, 1994.
6. K.S. Krishnaswami, Agriculture Development under the new economic region EPW. June 15, 1994.
7. Vikas N. Pandey, Agrarian Transformation and Co-operatives continuity and change EPW, June 25, 1994.
8. Jurgen Von Myralt, The role of co-operatives in the process of structural adjustment : New challenges and ILO's approach. Invited distinguished lecture at Symposium on Management of Rural Co-operatives, December 7-11, 1992, IRMA (Unpublished).
9. Mohanan, National workshop on co-operatives and new economic policy NIRD, 1993.
10. Mohanan, (I) Impact of new economic policy on agricultural co-operatives background paper (I) NIRD Workshop, September, 1993.
11. Dr. Jha, Challenges and prospects for co-operative development in India under new economic scenario. Background Paper II in NIRD Workshop, September, 1993.
12. T. Haque, New Economic Policy and Agricultural Co-operatives Background paper (#) NIRD, Workshop 1 September, 1993.
13. CDF, Deregulation and co-operatives Background No.(4), NIRD, Workshop September, 1993.
14. P.R. Dubhashi, Opening up the co-operatives, Financial Express, 3rd June, 1992.
15. P.R. Dubhashi, Co-operatives and New Economic Policy Financial Express : 17 Nov., 1994.
16. K.K. Taimini, Co-operatives and structural adjustment, Economic Times, 25 April, 1993.
17. G.R. Bhave, Co-operative Movement and Planned Economic Development in the context of New Economic Policy, ISCS, 1993.
18. R.V. Nadkarni, Strategies for meeting the challenges through upto

co-operatives in the wake of new economic policy of the Government of India, WAMNICOM, Pune 1993 (Republished by CDG NDDB, 1993).

19. Tushar Shah, The Logic of a co-operative company. The IRMA Symposium on Co-operatives, 1992.
20. S.N. Singh, Give Companies Status for Co-operatives NDDB, Anand.
21. S.N. Singh, Present crisis in Indian Co-operative Movement NDDB, October 3, 1989.
22. S.K. Saxena, Leadership in Co-operatives, Some observations Coop. Dialogue, Vol.3 No.1 and 2, 93 March, 94,
23. Daman Dharia, Co-operative Movement, Search for new direction V.M. Memorial Lecture, 1992, NCUI, Orissa.
25. V.Kurien, New Economic Policies and Co-operatives, 'Keynote Address, Bangalore, Organised by National Coop. Union and International Labour Laws (ILO) September 1.
26. Dilip Shah, Cooperativization in India : Need for a fresh and positive approach, Initiative and actions (An unpublished mimeograph).
27. K.V. Nambiar, Cooperative and Development, Financial Express.
28. E. Michael Johnson, Co-operative Public Companies, Economic Times, 5th July, 1993.
29. N.L. Vidyanathan, Lecture delivered during panel discussion on Co-operative Companies at IRMA, on 15.06.93 organised by working group on current issues at Anand.
30. S.R. Tripathy, Co-operative Societies Act. Changes in Step with time - Business Standard, Calcutta 1, August, 1993.
31. Kurien V., The Proposal for amendment of the Companies Act for facilitating establishment of Multi State Co-operative Companies. National Trust on Cooperatives at New Delhi on Friday 3, 1993.
32. V. Kurien, Parliamentary Standing Committee on Agriculture Presentation on National Agricultural Policy Resolution, September 1993.

□□□

3

Co-operative Legislation in the Context of Economic Liberalization

Introduction:

This brief paper highlights relevant facts and Issues concerning co-operative legislation and analyse them in the context of prevailing environment of economic liberalization adopted since 1991. The paper is designed to discuss briefly the following :

1. What is the role of co-operative legislation ?
2. What actually is the contribution of co-operative legislation in co-operative development ?
3. What exactly are the changes urgently required to make co-operative constructively instrumental for genuine development of co-operatives in India.
4. What is the status of steps being taken to change co-operative legislations in India ?
5. What could be done to accelerate and complete the process of changing co-operative legislations in India ?

Each of these question isdiscussed in brief in the specific section of the paper, then the paper ends with an appropriate concluding observations.

Role and Place of Co-operative Legislation in Co-operative Development

Though a lot of blame is drawn on account of co-operative legislation for the prevailing status of development of India. We should never undermine the role of co-operative legislation which has rightful place in the powerful people oriented genuine co-operative development. The following are some of the basic points indicating the real role of the co-operative legislation in the true sense of spirit :-

1. The co-operative law is required to ensure that co-operative enterprises are developed and grown in the basic framework of national constitution which is the basic law of the nation.

2. The co-operative laws are needed to clearly defined cooperation and should be in tune with basic ideals and principles of co-operatives. As rightly pointed out, "If co-operative legislation is to serve as an

instrument for upholding co-operative Ideology, it seems that the co-operative principles should find an explicit place in the body of the basic law, one possible approach is to see that the each accepted principle is clothed in legal sanction by one or more section of the Act, thus, as a whole will tend to ensure compliance with co-operative principles".(1) Any provision of co-operative legislations inconsistent with co-operative principle bound to develope institutions which are far from being co-operatives.

3. The co-operative legislation should be capable of promoting peoples Initiative and self-reliance.

CONTRIBUTION OF CO-OPERATIVE LEGISLATIONS IN CO-OPERATIVE MOVEMENT OF INDIA

The various body of the experts also do support the facts summarised by the various committees. We may quote a lengthy study by NDDB, the observations of 1987 and also Brahm Prakash Committee of 1991, Dr. S.N. Singh's study has very successfully shown how co-operative Acts not only "Co-operative" but also violated the principles of co-operatives(4 by killing the spirit of self-reliance. In the words of a bureaucratic Managing Director Jayprakash Narayan, Government on account of such co-operative Acts failed to provide good governance, got on the back of co-operatives and led too much discriminative policies in the name of protection.(5)

Finally, co-operative legislations should have adequate dynamism to meet the challenges of new economic policy i.e. Liberalization of economic policy.(6)

The committee on co-operative law of 1987 have made the following remarkable observations :

A. The existing co-operatives society Acts contain provisions which militate against the democratic character and the autonomy of co-operatives.
B. Over the years, the Registrar has acquired undue power in respect of management.
C. The role of the Registrar was not positive.
D. The federal co-operative did not play active part in development.
E. The changing complexities of management; human resources development, financial management have necessitated availability of professional assistance which was not made available to co-operatives.

1. The Brahm Prakash Committee has listed the following Restrictive Provisions, now in force empowering Registrar to be a creator, preserver and destroyer of co-operatives rather than becoming Friend, Philosopher and Guide.(7) Compulsory Amendment of By-laws by the Registrar.

2. The power of Government to nominate directors on the committee of management.
3. Power of Government to veto annual/**Rescind** resolutions.
4. Powers of Registrar State Government to give directives.
5. Supersession and suspension of committee of management.
6. Automatic supersession of committee of management of co-operative credit societies.
7. Restriction on terms of office bearers of co-operative bearers of co-operative societies.
8. Restriction on holding office in a member of co-operatives simultaneously.
9. Compulsory Amalgamation and division of society by the Registrar.

Implications of all these provisions are so obvious that it requires no analysis, however, it may be illustrated by the fact that even the high court do not endorse such provisions. For example, Madras high court has struck down the Tamilnadu Government Action which suspended about 27,000 co-operative societies and vested their management with special officers with one stroke of pen.(8) The International opinion has also gone against such provisions, for example, after due discussions in conference of co-operative Ministers organised by ICA at Sydney (Australia) in 1990, the "Sydney Declaration" passed has clearly stated that "Co-operative by nature are autonomous institutions. Government should play the role of facilitator through policy initiatives and should nurse and nurture them to play the full role as people's organizations". The Sydney Declaration suggests that "Each member country should enact progressive co-operative legislation in conformity with co-operative principles, and take steps for deofficialization of co-operatives".(9)

Thus, the present co-operative Act did not serve the objective of developing genuine co-operatives. In fact, those who really want to develop co-operatives in the right sense of the term says "For heavens sake, give us good Governance, get off our backs and give no discrimination treatment" by changing the co-operative legislation.

Finally, the present co-operative legislation has failed to serve the dynamic needs of the time. Economic liberalization has opened up the great opportunities for development of co-operatives in variety of ways, however, the chances of realizing this potential is relatively blank because of the strings of co-operatives acts which were passed to serve the colonial need of British Government and further strengthened by the politicians to preserve their power after an independence. This observation has a support of the fact that, the best possible model act drafted by the planning commission appointed by Brahm Prakash Committee had been strongly opposed by political leaders.

EXPECTED CHANGE IN CO-OPERATIVE LEGISLATION

The process of amendments is so large that it calls for almost new enactment of the state legislation as well as multi Co-operative Act legislation at the central level. It should reverse concentration of powers to state. It should drastically change the mode and approach from colonial character to norms of governance of peoples.[1] Autonomous movement the co-operative legislation should clearly defined co-operatives and framed itself in terms of ICA principles and redefine the role of registrar and role of states. It should have enough provisions to make the challenges of economic liberalization faced by the co-operatives. For example, delicensing dairy manufacturing has opened the "Competition" between co-operatives and non co-operatives. In that situation, obviously, co-operatives should have enough opportunities to contest at least on equal footing. For example Dr. Kurien has very excellently summarised the needs in the 'Five decisive steps' urged to the Prime Minister to energise India's rural co-operative.[(10)] Accordingly, these steps are:

1. A New law for co-operatives.
2. Allow co-operative companies.
3. Implement new labour law.
4. Rationalise co-operative taxation.
5. Get the Bureaucracy out of co-operatives.

These five steps are most vital to energise the co-operatives, to serve the rural people against the adverse impact of new economic policy, however, the substantial changes in the co-operative legislation is the first and foremost step for developing co-operatives which is self reliant and responding instrument in the hand of rural people.

IV. STATUS OF STEPS BEING TAKEN TO CHANGE CO-OPERATIVE LEGISLATION

Various committee on co-operatives right from 1954, All India Credit survey to Mirgha Committee on co-operative law in 1967 attempted to give new orientation to co-operative laws and infact, Ardhanareswaran Committee on co-operative law 1987 made very harsh observations regarding excessive concentration and restrictive measures, however, neither state governments bothered to change the law nor these committee could provide 'Comprehensive Model' however, it was in March 1990, the Planning Commission appointed the Brahmprakash Committee which not only thoroughly reviewed the state status of the co-operative but also provided the Model Co-operative Act", published by Planning Commission in May, 1991 is so comprehensive that more or less its adoption by the states could serve major purpose and eliminate the most of the loop holes of the laws leading to hindering the real growth of co-operative which enable them to take advantage of economic liberalization.

The approach of the model act is two fold i.e. to provide a genuine character to co-operatives and to facilitate building of an integlutade co-operative structurel so as to evolve a co-operative system. The special features of the model co-operative act which are summarised by the committee are as follow.(11)

1. The Act provides for the state policy, Principles of co-operatives and also relevant definitions including definition of co-operatives as a guide to confirm the basic ideology of co-operation by state.
2. Procedure for Registration of a new co-operative is simplified.
3. The Model Act gives rule making power to the Government expecting broad parameters to be observed by co-operatives in framing the buy-laws etc.
4. The model Act provides no power to the Registrar or government for any of the following such as :

a) Suppression of the Board of Directors.
b) Compulsory amalgamation or division of societies.
c) Compulsory amendment of the by-laws. escind/annual the resolution.
d) Issue Directives.

V. Co-operative Federation/Unions are to assume greater responsibilities towards the member co-operatives relating to election, and audit etc.

VI. The role of Registrar under the model Act has been confined to the registration and liquidation of co-operatives, conduct elections, audit and to convene meeting of general body.

VII. The Model Act prohibits co-operative from accepting funds from the Government by the way of Equity.

VIII. Special obligation have been imposed on members to ensure the character of co-operative as a member user organization.

IX. Board of Directors are made accountable for timely election, audit, regular meetings etc.

X. It prohibits officers of the Government to work on co-operatives rescind and take cognizance of any offense under Act.

XI. Finally, the Model Act provides for the constitution of Co-operative Tribunal for the settlement of disputes including matter relating to constitution of management and Business of a co-operatives.

Thus, the model Act provides an amendment of the multistate co-operative Act in conformity with Model Act remains to be done. Infact, even after doing that, we shall need to do some amendment in the companies Act also to serve the needs arising out of these transaction of business among co-operative whose membership resides in more than one state. The multistate co-operative act does not extend to these co-operatives unions and federations whose business is across the states and even nations due to exports. Therefore, only amendment in the company Act could complete the task of providing adequate legal set up to co-operatives in liberalised economy.

However, the acceptance and adoption of the model Act is not encouraging. For example, when the ministers of Agriculture called a meeting of the state ministers and secretaries of cooperation, each and every state express its "Opposition" to the Central provisions of the model Act.(12) However, a co-operative initiative project headed by a panel of stalwarts like Mohan Dharia, Laxmi Jain and Dr. V. Kurien brought out their first status report (1993/94) on state of co-operation in India, which has observed the progress made in the adoption of Model co-operative Act.(13) It has noted that the standing committee on agriculture has recognised the need for liberalisation of co-operative law to enable co-operatives to plan and formulate their activities. Mr. Balram Jakhar, Minister of Agricultural has also made a public committee to amend-? the multi-state co-operative societies act on the line of the model. Act and promise to support amendment of the companies Act to enable in Corporation of multistate co-operative companies.

Moreover, some state Governments such as Uttar Pradesh, Orissa, Madhya Pradesh and Karnataka have expressed their willingness to amend their state co-operative laws in the line with the Model Act, so far Gujarat state has concerned its 1961 Act was first amended in 1981 through an ordinance which was rectified in 1982, however, as documented and analysed by experts, the whole process was purely "political" aiming at cutting the political power of each others in caste based politics of Gujarat and it was never an exercise in the line of Model Act.(14) Thus, one of the most progressive state in cooperation is also far behind in accepting and adopting the model Act.

Thus, the present status of movement of changing co-operative legislation is very discouraging and extremely uncertain and terribly slow.

V. SUGGESTED ACTIONS

The status report of co-operative initiative project (NDDB) has documented many facts showing very negative steps and expression of the same attitude of the ruling classes which has hampered the growth of real co-operative movement.[15] Therefore, what it broadly suggests that "If the full potential of the co-operative movement is to be realized, all concerned must without delay, strive to purse the environment of those elements that stifle the movement and deprive it of its burnt vitality". This sentiment could be expressed in terms of specific, actions to be persuaded by concerned agency and organization. Briefly, we may summarise the main actions suggested by the co-operative initiative project which has undertaken the national task of developing vibrant co-operative sector in India.

- All political parties must incorporate strong statements in their election manifestos in favour of the amendments of co-operative legislation in the states and Union.
- The co-operative leaders and practitioners should educate people and encourage national debate on the issue.
- Infact, the awareness of the people on the issue is too less even to understand the real strength of the proposal.
- Academic class of experts i.e. Teachers and researchers who believe to develop strong people's economic institutions like co-operatives must come forward for independent studies, persuasive papers and even popular writings in the news papers to educate people regarding the issue and help building positive mass opinion.
- The co-operatives themselves must undertake a major task of education and explaining its members regarding the significance of this issue and prepare them to work as carrier of the change in the rural areas.
- Finally, NGO's have to play a vital role in taking a lead at grass root level to magnify the major aspects of the co-operative legislations both current one and model Act so that a public opinion could be built up to compel the states to undertake right steps in right time ultimately adopting and replacing the current act by modern model Act which has potential of revolutionary change.

CONCLUDING OBSERVATION

Anti Co-operative legislation obviously could not be the purpose of meeting the needs of liberalization of co-operatives. To eliminate the damages done by the act in the development of real and robust co-operatives, we must endorse the agenda of actions proposed by the co-operative initiative group and do our best to implement in such a manner that we can secure sufficient political pressure for the earliest acceptance, adoption and enactment of co-operative societies Act at the all India level in the line of model co-operative development in India very much depends upon the extent of the success in changing co-operative legislation right in time.

REFERENCES :

1. Planning Commission, Government of India Report of Brahm Prakash Committee on model co-operative Act, New Delhi, 1991 pp. 415.
2. S.N. Sinha and Rama Sethy Ibid pp. 156.
3. Co-operative Initiative Project :
   State of Cooperation in India, Comparative year 1994. Co-operative Initiative Project IRMA campus, Anand.
4. S.J. Phansalkar
   Co-operative in the policy of Gujarat state in Workshop on state and co-operative in India, Symposium on management of Rural Co-operatives, IRMA, 1991, pp. 24.
5. Co-operative initiative Group. "state of cooperation in India." (1995) N.D.D. & ANAND 1995.
6. V.K. Kurien : Dive decisive Steps : Note to Prime Minister N.D.. 1992. Five
7. of India : Planning Commission Report on the Model Cooperative A. New Delhi, 1991.
8. S.N. Sinha and Ram Sahy Ibid, pp. 1511.
9. Cooperative Intitive Project : State Cooperation in India & Cooperative year 1993-94 - Anand IRMA Campus.

4

Co-operative and rural Development in the context of Environment of Economic liberalization

INTRODUCTION :

Development through the Co-operativization is an attempt to seek "Middle path" and India has endorsed it against the collapse of the Soviety style socialism and increasring evidence of market socialism in many countries of the Africa and latin America. (The remarkable success of co-operatives in some regions in India confined to some sectors and also in some parts of developing countries has provided new vision and revived the optimism that co-operatives can provide an alternative as "Middle path" incorporating development interest of poors also. As it is capable of combining the efficiency and productivity of market friendly liberalization and with socialist concern of welfare poor.

It is within this broader prospective we have designed this brief paper to discuss the issues of rural development in the context of the environment of economic liberalization and feasible solutions offered by the robust design oriented genuine co-operative organizations. The paper is divided in terms of following sections :

(I) Fist section deals with the concept of rural development and lists out the issues which are either slowering down or leading to failures in achieving rural development beyond government targets of financial and physical success.

(II) The second section briefly review the quantitative growth and structural change of co-operative sector. Successful models of co-operatives provide an evidence of the co-operativization as an instrument of rural development.

(III) Third section critically examines the constraints coming in the way of realization of such potential particularly in the context of the economic liberalization since 1991.

(IV) Fourth section briefly prescribes specific policy actions to develop efficient co-operative sector which also serve the rural producer who are also mostly in the class of rural poor.

(V) The paper ends with some very specific observations to indicate the direction of efforts required from all those who are really co-operative friendly and also rural poor friendly.

I. RURAL DEVELOPMENT AND RELATED ISSUES

To survey the actual issues in rural development we will have to appreciate the fact that there are two different conceptualization of rural development in theoretical literature and at policy level in India. We have dealt with this point in detail elsewhere, it will be sufficient to note in this brief paper that Rural development envisaged in theory is quite different than followed in policy. Theoretically, Rural development implies a simultaneous development of agriculture as well as non-agricultural sectors of rural area with active and continuous participations of rural poor with state intervention, and assistance leading to eradication of their poverty : Promoting employment and reducing inequality on permanent basis. Against this what we find in the development plan of India in the name of Rural development is an adhoc intervention of state with anti-poverty programmes unrelated and unintegrated to total plan. In fact, planning Commission equates anti-poverty programmes with rural development. Though, we do not underrate the significance of this 'Growth plus strategy as otherwise it would have been impossible to alleviate the poverty even by 16% rate of growth as calculated by C.H.H. Rao as an alternative to anti-poverty programmes.

However, we should list the issues related to rural development touching upon both the version of rural development on the basis of variety of critical studies and evaluation studies we may briefly state the following issues :-

(I) Despite the recent positive effect of Anti-poverty Programmes via reduction in consumption in equality on reduction of poverty ratio; in terms of coverage and also its permanent impact, it is much inadequate as hardly 15% of rural poor are benefited and those who cross poverty line needs either second dose of assistance or they go back to below poverty line. The point is to empower poor adequately and in certain way so that they themselves could take care of their problems.

(II) A large number of RDP fails due to absence of active and meaningful participation in their development.

(III) As happened in IRDP related dairy finance, their investment without organization failed to alleviate poverty and beneficiaries boxed in debt traps, however, where active co-operatives supported, like that of Sabarkantha they did a miracle as demonstrated in a study of Katarsing.

(IV) Sincerely followed programme DWAKRA faced the crisis as there is not market linkage of such programmes.

(V) Excellent technological support did not click in the case of upgradation of traditional skill and technology due to non-availability to market.

(VI) Every effort to promote non-farm economic activities through IRDP failed due to poor scales, poor bargaining power, and non-available infrastructure, processing and modern manufacturing for value, addition etc. particularly on account of poor resource base of the rural poor.

(VII) The means of enhancing opportunities to develop co-operatives to Fight against 'poverty' has gone on declining due to liberalization in the area of health and PDS etc.

(VIII) The equitable access to common pooled resources like water, waste land, forestry etc. which had played positive roles in sustaining weaker sections has declined substantially for several reasons including encroachment by rich and controlled regulation by state on environmental grounds.

(IX) Institutional credit has not slown adequately to support weaker sections including SC/ST Classes and recent financial reforms also affect adversely due to redesigned priority sector and revision of interest rate policy and limiting the role of PSCS including RRBS etc.

(X) Rural development did not really clicked as the best stated by Dantwala who says A direct attack on rural poverty without equally direct attack on the socio-economic structure : C.T. Kurien has rightly asked : "How could we correct the consequences or themselves without curing the characteristics.

II. RURAL DEVELOPMENT STRENGTHENING GROWTH OF CO-OPERATIVE SECTOR AND SEVEN SUCCESS STORIES OF CO-OPERATIVES DIRECTLY RESOLVING RURAL DEVELOPMENT ISSUES :-

The co-operative sector developed in rural India is largest next to China. In recent times, it has not remained only 'credit' but also developed as non-credit business and thus assuming a vital role of "Agro-Rural Business" touches upon 2.13 lakh co-operatives of 15 different types having 847.67 million members spreaded all over India which deal with Rs. 868,157 million working capital. Its 'Third sector' significance has felt clearly in rural economic development as 43% of total agricultural credit, and 34% of fertilizers are distributed by co-operatives and thus, became carrier of the green revolution. Even in small scale sector, as much as 58% handlooms are owned by the Co-operatives. A table in Appendix provides more detailed scenario of quantitative spreaded co-operatives in India.

However, much more significant fact is that during last 50 years number of successful co-operatives have emerged as "MODEL" co-operatives covering different areas and different items which are directly related to development of rural poor and rural development. We may briefly take note of such seven success stories.

(1) ANAND PATTERN DAIRY CO-OPERATIVE SOCIETY (APDC):

APDC models needs no introduction in India due to successful three phases of operation flood projects. As of March, 1996, 72000 APDC operate with over 9 - 10 million producers spreaded over 170 milksheds of India. According to NCAER it pumps Rs. 30,000 crores everyday in rural India. It has proved beyond doubt how it promotes income, employment and standard of living and generate 14.18 crores of per day employment. In fact, it is better and more successful rural development project than Government IRDP/ Anti-poverty programmes. Anand experiment became Anand pattern by successful replication in Gujarat and now it has become a 'Model' due to replication in non-Gujarat states in larger scale.

(2) TRIBHOVANDAS FOUNDATION RURAL DEVELOPMENT PROJECT (TFR) :

TFRP is an excellent extended model of rural development of Anand pattern dairy co-operatives. An integrated rural development self-reliant project is well designed and implemented using well spreaded dairy co-operative network, all over the Kaira district, we have elsewhere evaluated the project, however, it may be stated that it has done miracle in improving health and nutritional status of children and expected women.

(3) SUGAR CO-OPERATIVE MILLS OF SOUTH GUJARAT AND MAHARASHTRA (SCH) :

Rural prosperity of South Gujarat and Maharashtra is greatly attributed to the successful members' controlled, members' interest protecting sugar co-operatives. Tushaar Shah and Baviskar both who have different perception of co-operatives have equally admired sugar co-operatives off course for two different reasons. For example, Baviskar attributes success of sugar co-operatives to stable alliance between larger farmers controlling sugar co-operative management and small and medium farmers who are sugarcane suppliers, forced by need of critical viability of sugar co-operatives. Tushaar Shah has attributed the success to the members own control.

(4) **ANAND FARMERS SERVICE CO-OPERATIVE BANK (FSB) : MODEL IN CO-OPERATIVE RURAL BANKING :**
This promising model of Rural co-operative Banking is complete multi-service model developed in Kerala is an innovative model in the sense that it has not only financial bank branches but also farm input depots, agricultural commodity, procurement centres and consumers stores. This real co-operative in terms of spirit and action is emerged within the same macro-credit policies structure which are being questioned these days. Bhupat Desai has assessed its viability in terms of its profitabilities, loan delinquency and scale and scope economics in various costs.

(5) **MOHINI WATER CO-OPERATIVES MODEL :**
As Robert chambers had once pointed out that irrigation access is one of the greatest source of empowering poor and ensured some of small and marginal and landless against monopoly of water Lords through water market is best illustrated by "Mohini" water co-operative society as early as 1978 such irrigation co-operative in canal command area of Sardar Sarovar Project are extremely important to extend benefits to small and marginal producers in command area.

(6) **GAMBHIRA COLLECTIVE CO-OPERATIVE FARMING SOCIETY MODEL :**
The Gambhira Collective co-operative society which emerged to develop flood affected people and waste land longback in 1980 provides an illustrative model how marginal and landless poor could be provided with 'surplus land' due to land ceiling and 'Waste Land' alongwith suitable institutional support for viable farming through pooling of limited resources and state support. Gambhira also provide an example of integrating private incentives with collective strength for the promotion of common goods and welfare.

(7) **TREE GROWERS' CO-OPERATIVES SOCIETY (TGCS) MODEL OF NDDB :**
Role of co-operatives in natural resources management in promoting sustainable development and rural poor is a matter of recent phenomena. Recently, developed TGCS are providing

unique possibility of what Swaminathan once suggested a type of rural development which can be poor friendly and environment friendly as it ensures both simultaneously flood security as well as environmental security. The institutional structure of this kind has already been established in 5 states of India i.e. Gujarat, Andhra Pradesh, Orissa, Karnataka and Rajasthan. TGCS model aims at motivating its members and other village people to grow trees and grasses of suitable species on their own marginal agricultural land and forest lands to meet the local needs of fuel wood, fodder and small timber and to provide livelihood by improving quality of local environment. The TGCS holds high promise as an instrument of restructuring forestry in India on a sustainable basis involving rural poor.

Thus, quantitative spread and structural changes in co-operative development alongwith specific models of co-operative development hold much greater potential and promise to meaningfully promotion rural development on a sustained basis with participation of those who are otherwise marginalised from main market dominated development trend and the development process in the India economy.

III. CRITICAL QUESTIONS CONCERNING CO-OPERATIVIZATION IN INDIA :

In our earlier sections, we could draw a picture which provide the hope to develop institutional instrument for rural development focussing on rural poor, however, there are very specific critical questions which have emerged out of Indian experience of co-perativization and its theoretical and empirical direction. We have to understand, appreciate and resolve these questions successfully to make co-operatives real instruments of development of rural poor. Though we have analysed these questions in detailed elsewhere, we would briefly state these questions for the purpose of raising debate on these issues. These questions are as follow :-

1. **Objectives and operational difficulties of co-operation :-** Objectives of co-operatives are well guided by principles of co-operatives, however, there are now quite powerful arguments leading to explain that some of the basic difficulties of survival of co-operatives come very much from dogmatic operation of such principles. The review of principles of objectives called for also the economic liberalisation entering through delicensing dairy industry.

2. **Analysis of chemistry of success of the success stories and models of co-operatives.**

Recent literature in co-operative development has made an interesting attempt to explain the factors behind the successful co-operatives. For example, Baviskar Atwood Thesis relating to social structure, stable alliance between different class of farmer and strong sustainable leadership is quite convincing but Tushaar Shah has rightly criticised it. It fails to provide explanation for all types of co-operatives including low performing co-operatives. Against this inadequate explanation, we find a theoretical framework analysing the internal dynamics of co-operative performance bringing out some critical elements of success leading Chemistry of success in Tushaar Shah's studies. These elements are :

1) Strong drive to success, strong collective will for the success.
2) Member first model over value first model
3) Service to members : utility for members to each of these patronage, a real strength and source of success
4) Social commitment
5) Robust design concept.

These factors are more or less present or absent in successful or failed cooperative societies.

3. **Replication or promotion of successful co-operatives :**

The question of multiplying many more Anand pattern is as old as operation flood projects (1970). Recently, this question has been readdressed more effectively in terms of Golden Hydens' Blue print - Green house approaches which were applied to explain African's experience. Blue print approach implies a notion that tested model exists which can be applied and reflected in an effort at planned development. The green house approach is based on the assumption of the 'slight stimuli' and 'incentive' which would be capable enough to generate an organization which will accomplish task of common interest, while both the approaches have their own limitations, one would favour the possibility of developing successful co-operative model through Blue print approach, provided quality of original design concept, quality of replication process and adequacy of minor adoptions are available.

4. **Distributional dimension of developing co-operatives :**

Since rural development by concept is based towards rural poor, who will get the benefits of rural co-operative development, is a moot question, however, this question has generated more heat than light as it has not been seen in the proper perspective. On the one extreme, it is believed that the objective of equality is super imposed normative value. On the other hand, it is believed to have developed as an instrument of 'co-operative-capitalism', however, careful analysis of sugar and milk co-operatives sector shows that though may not in equal proportion; small and marginal producers did have got benefitted from co-operatives, however to maximise the active participation of resource less poor in rural co-operatives is a pertinent question associated with effectiveness of co-operatives in developing rural poor and development.

5. **Review of the role of the state in Co-operative Development and Management :**

The excessive role of the state has hampered the natural and efficient development of co-operatives. Its role is not only very regulatory but dependency generating and developing clutches of bureaucracy and politisation of the co-operative movement which has ultimately what V. Kurien says "hijacked the whole movement for the narrow and nasty benefits of political leaders." The concentration of decision making power is so much with the registrar that registrar of the society is not facilitate but father of the whole system. State has indeed protected and encouraged unreal, enviable and even ghost co-operatives for variety of vested interest Corrupt co-operative leadership is even characterised by contrary values to co-operative principles.

6. **Restructuring and Redisovering of co-operatives in an environment of economic liberalisation :**

Economic liberalization led deregulation of dairy sector and thus opened up the challenging issues of survival and development for the co-operative sector. Co-operatives will have to be restructured in terms of their status, size, style and methods to adjust with challenges of competition from private sector, withdrawal of the state financial support, and withdrawal protective/reservation policy shelter and even inter-co-operative competition for market. Thus, these critical issues will have to be resolved to develop strong sustainable middle sector of the economy promoting rural development.

IV. SPECIFIC POLICY ACTIONS FOR POWERFUL RURAL CO-OPERATIVE DEVELOPMENT :

In the light of the analysis of the main issues faced by the Indian co-operative sector, we may suggest following major policy actions to redesign co-operative sector with new strength and efficiency to play vital role in the promotion of rural development. Some of the major policy actions are as follow :

1) **National Co-operative Policy :-**
   It is surprising and shocking that India has no national policy. The specific national policy statement spelling out specific objectives, approach, model of co-operative structure, major instruments and specific tasks along with clear role of the state could be very helpful to truly liberalised co-operatives to take advantage of new economic policies.

2) **National initiative for co-operative Development :-**
   National initiative with powerful public movement reasserting role of co-operative movement and calling for the essential changes in the co-operative policy leading to revitalize the co-operative is essential. N.D.D.B.'s co-operative initiative project in collaboration with NGO has provided a model in this direction. This project should be regionally spreaded with state level penals of respected regional experts and leaders like V. Kurien L.C. Jain and Mohan Dharia.

3) **Promotion of Members' participation :-**
   The greatest strength of any successful co-operative next to peoples' own initiative is members' participation in co-operative control. Its the key to promote member First co-operatives. This could ensured only by tangible economic benefits and education to member about the co-operations. Co-operatives education make them 'aware' of their duties and rights and help them for participation which is very useful to develop efficient co-operatives. Adequate environment also has to be provided to members to adopt what they learn in such programmes.

4) **Abolition of antiquated co-operative societies Acts and Adoption of the Model Act of Co-operatives :-**
   The co-operatives need bold steps to reform and restructure which calls for immediate abolition of an antiquated, colonial co-operative Act. it should aim at clearly spelling out the role of the government such as incentives, inspection and audit. The

planning commission has already developed such 'Model Act' however, it has not been adopted by any state except Andhra Pradesh.

5) **Development of strong Federation union :-**
The model act has rightly provided a great role and power to federation/unions. As rightly stated by HB Shiva Maggi, Indian co-operatives have a paradox of strong federation with sick and stagnant village co-operative societies which could be explained in terms of absence of active and positive responsibilities of federation towards lower level societies.

6) **Constitution of Co-operative Tribunals for corrupt practices, co-operative crimes and adjudication of disputes including appeals under the Act :**
The most important are the co-operative crimes in the form of misusing funds and corrupt practices which have tarnished the image of the co-operatives among public. The regular judicial procedures are lengthy, complex and time spending, hence to correct this source of weakness an idea of constitution of co-operative tribunals of disputes must be extended further. Such judicial co-operatives should be available to every district with judicial offices in the rank of district court judge.

7) **New provisions to encourage co-operatives for the maximum benefits or liberalization policies of trade and financial and international trade :**
The new economic policy of restructuring and stabilization has certainly open up opportunities for development on the part of those non-credit co-operatives which are as large as any private or multi-national firms. They not only need equal playing ground to compete with privates and multi-national but also appropriate alternative avenues arising out of withdrawal of state and abolisation of restrictive, regulative and resolving policies adopted to 1991.

To develop environment for really liberalised economy state must make provision for the following :

1) Strict enforcement of milk and product order.
2) Amendment to permit co-operative company.
3) Tax reforms to eliminate double taxation and over taxation.
4) Labour law reforms to make labour law provision appropriate to co-operative units.
5) Withdrawal of upgradation on distribution of profits etc.

8) **Development of Co-operative Professionalism :**
The nature and size of Co-operative development in India denies now simple political ordering and trading methods of managing co-operative sector. Katar Singh has brilliantly pointed out the case of professionalizing the management of Agri-business co-operatives. It is correctly argued that 'agri-business co-operatives need to be managed by professional managers. If they are to survive and grow in the wake of India's new economic policy characterised by deregulation, delicensing privatization and globalization. However, apart from the development of Management culture among co-operative, there are several obstacles to professionalise Management.

9) **Strategic Policy actions for viability of the co-operatives :**
The unviable and inefficient co-operatives can not provide alternative to private or public sector nor it could become a powerful developmental instrument for the rural development. The economic tangles leading to unviability such as higher unit costs, under utilization of capacity, higher transaction and service charges, losses, leakages, overdues, below cost pricing, narrow margins, poor own funds, etc. will have to be resolved in the light of various recommendations of expert committee and other research studies such as Khurow committee and others. Bold initiative and efforts will be required to introduce "restructuring process" with a variety of package of actions suitable to different situations. Such strategic action may include tied contracts, amalgamation, reorganization of units, strong support from federation, quality control measures, new market opportunities drive, appropriate technology, diversification, rotation system of internal control and incentives and promotion of internal funds etc.

10) **Participation of poor in Co-operative Management :**
Finally, no miracle is likely to happen to involve poor in production process : the deliberate efforts to involve producers in the compass of co-operatives perhaps would be helpful to achieve this : unless this happens co-operatives just could not do anything to contribute in development of rural poor. (Apart from variety of moels suggested. We may implement with adequate modification, we should also be able to revive traditional co-operatives on the line of suitable traditional models by Anil Gupta and Umalele etc.)

CONCLUDING OBSERVATIONS :-

It is well established that co-operatives could be a great institutional arrangement to promote rural development in the real sense, however, the present scenario does provide a remarkable ray of hope, in the light of economic liberalization dominated environment, unless co-operatives are not truly liberalised, we shall perhaps miss a great chance of developing a "middle path" and shall be left with showpeace of models having no relevance in resolving rural development crisis which will take its worst form in the next phase of economic liberalization.

REFERENCE

1. **Dilip Shah**
   Critical Review of Rural Development in VIIIth Plan **Financial Express.**
2. **B.S. Baviskar and Donald Attwood**
   Finding the middle path - The political economic of Co-operatives in Rural India. Chapter 1, View, sage, 1996.
3. **Tushad Shah**
   Emerging India's Rural Co-operatives Draft paper, Circulated IRMA, 1991.
4. **Gorden Hyden**
   Approaches to Co-operative Development Blue print Vs. green house in Atwood-Baviskar edited who shares ? Co-operatives and Rural Development, Oxford, 1988.
5. **Katar Singh**
   Some contemporary issues in professionalizing the management of Agri Business Co-operatives in India with special reference to Dairy Co-operatives, W.P. 102, IRMA, July, 1996.
6. **Katar Singh and Vishwa Vallabha**
   Role of leadership in co-operative Management on natural common post resources - A collective Goods Theoretical perspective. W.P. No. 50, Feb. 1994.
7. **Katar Singh**
   Role of Tea Grower's co-operatives in promoting Farm Forestry in India, W.P. 58, IRMA May, 1994.
8. **Tushar Shah**
   Re-engineering India's Co-operatives A Frame work for Actions, W.P. 80, IRMA, 1995.
9. **Vikas Pandey**
   Agricultural Transformation and Co-operatives : continuity and change, EPW, April, 1994.

10. **Dilip Shah**
Rural Development through co-operative approach in D.R. Shah (edited) alternatives in Rural Development.
11. **H.B. Shivamaggi**
Future strategy for development of co-operatives EPW, May, 1996.
12. **Dilip Shah**
Co-operativization in India : Need for a Fresh and positive Approach, Initiative and Actions (unpublished paper).
13. **Dilip Shah**
Role and Future of Co-operative Management in Das (edited) Co-operatives and Economic Development Mittal, 1996 (Forthcoming).
14. **Dilip Shah**
Co-operative legislation in the context of Economic liberalization in Das (edited) co-operatives and Economic Development, Mittal, 1996 (Forthcoming).
15. **Dilip Shah**
Co-operative reforms in Gujarat **Financial Express** 12th June, 1996.

APPENDIX

A profile of Co-operatives in India (as on March 31, 1993)

	Type of Co-operatives	Number	Member-ship	Paid-up Capital	Reserves	Working Capital	Business Operations
				Rs. Million			
1.0	**Credit Co-operatives**						
	1. State Co-op. Banks	28	98	28	108	1,56,456	59,977*
	2. Central Co-operatives Banks	352	1128	10684	11776	2,43,907	88,504
	3. Central Land Development Banks	19	2745	3296	3978	64,653	11,633
	4. Primary Agricultural credit societies including LMPS & FSS.	91210	93610	15728	5388	1,35,950	70,623
	5. Primary Co-op. Bank including Urban Banks	1326	15861	4778	10034	1,37,354	80,929
	6. Primary Land Development Banks	878	6415	35740	648	26,064	4,228
2.0	**Non-credit Co-operatives**						
	1. Marketing societies	8364	4869	1424	-	9,012	43,875
	2. Agricultural processing societies	1436	707354	336	-	1,946	814
	3. Dairy Co-operatives	73000	9263	4553	-	6,455	42,184
	4. Poultry societies	3218	202	57	-	317	270
	5. Fisheries societies	7189	903	322	-	1,010	1,536
	6. Farming societies	6370	343	69	-	443	215
	7. Sugar Factory societies	280	2618	10147	-	64,338	45,838
	8. Co-operative Opening Mills	548	558	3492	-	15,863	7,857
	9. Weavers' societies	17683	1707	1242	-	5,865	7,855
	All	212399	84674	91896	-	8,68,157	4,65,598

Source : - Adapted from Katarsing. Ibid. 1996.

□□□

5

Delicensing dairy products manufacture : Logic & reality

New Industrial Policy as an instrument of dynamic growth announced on July 24, 1991 is a part of the package of economic reforms and one of the most fundamental and far reaching political change. It abolished the practice of licensing industries under the Industrial Development and Resolution Act of 1951 and abolished licensing for all industries except for a short list of 18 industries which included the "Sugar" industry but excluded "Dairy products manufacturing". This policy of delicensing is a fundamental change in our national policy of developing dairy industry through small and marginal farmers and landless labourers by replication of the Anand pattern of dairy co-operatives attempted through the Operation Flood Projects. The merits and wisdom behind the dairy cooperativization are so well established that we may just draw attention to some of the latest studies which have again provided fresh evidence in favour of the strategy of co-operative dairy development in India.

LOGIC OF THE DELICENSING OF DAIRY PRODUCT MANUFACTURING:

It is now an open secret that the new economic policy (NEP) adopted by the present government is guided by "the conditionalities" imposed by the international financial agencies like IMF and the World Bank and these institutions are parts of the pressure tactics of the USA and other major developed countries. The "DUNKEL" Draft in the GATT is the latest imposition on India and other developing countries by USA. The New Industrial Policy is an outcome of this foundation. However, in the analysis of the logic behind the delicensing of the dairy sector. We may have to take note of the EEC policy towards the Indian dairy sector. It is a fairly known fact that a modest attempt by India for marginal exports out of the surplus production in 1988 had been responded by EEC by excessive reaction in suspending contractual dairy aid to Operation Flood-III and asking for an unscheduled review of the project. EEC is fully capable of influencing the international financial agencies to insist on liberalization by full entry into the dairy product manufacturing sector. This also could help to understand why 'Sugar' was excluded and 'dairy' was included in the list of liberated industries even though both belong to the cooperative sector in India.

The New Industrial Policy objectives could be the root of the logic of the NEP. It seems that these are five elements in the logic behind the NEP. These elements are listed as follows :

1. COMPETITION : It is expected to abolish 'monopoly' and to promote efficiency by encouraging competition.

2. INVESTMENT : It intends to promote investment particularly, foreign investment to promote growth and modernization.

3. GROWTH OF OUTPUT : It is expected that a liberal policy will further industrial production on account of the free flow of investment.

4. TECHNOLOGY PROMOTION : Development of indigenous technology through greater investment in R/D and bringing in new technology to help Indian manufacturing units to attain world standards.

5. EXPORT PROMOTION : The New Industrial policy emphasises on competition and welcomes Multi National Competitions (MNC) to promote exports by linking Indian manufacturing sectors to global markets so that we can acquire the ability to pay for imports and to make ourselves less dependent on aid.

We may now relate these elements of the logic to the Dairy sector in India today and try to examine its validity and relevance.

(1) COMPETITION : The competition could work as antidote to monopoly and this is in line with the NEP in so far as it wants to abolish the monopoly of any sector in any field of manufacture except on strategic or security grounds. However, the question is "what is the market structure of the dairy industry in India"? Liquid Milk marketing is always open and the co-operatives hardly handle 7% of the milk. Until recently major manufacturing unit has even taken interest in dealing with liquid milk distribution as it is a low profit product and even a loosing proposition. While huge profits could be made in dairy products. Before we criticize the milk business of the co-operatives, we should not forget that a well developed National Milk Grid System supported by adequate institutional and technological foundation has helped both consumers of the urban area and producers of long distance villages by developing a cross network of deficit urban area with surplus generating rural area which assured the supply and eliminated import dependence since 1975. However, due to unprecedented prolonged drought for three consecutive years some imports became inevitable between -1987 to 1989. The establishment of the Federations and introduction of infrastructural facilities are inevitable in such phenomena.

This strategy had become successful and was getting consolidated all over India. Under these circumstances, it is difficult to understand why one would like to delicense the dairy product manufacturers. Those who allege that the 'cooperatives' having virtual monopoly in dairy product manufacture should try to understand why dairy products are manufactured by Co-operatives and what role product these play in the economic viability of the co-operative dairy sector? The 'dairy products' are never produced by the co-operatives at the cost of the consumers' liquid milk. It generally follows the policy of producing dairy products in the form of powder which could be utilized to adjust lean period deficit by production of recombined milk and it is this phenomena which had made us independent of commercial imports. Moreover, the relatively better margin in the manufacturing of the product does help the dairy to balance their losses on account of the trade in liquid milk and in providing better procurement price close to the cost of production at the farm level.

It is also argued that competition will help producers as the demand for milk will increase. First of all it may be noted that we have no problem of demand. In fact, we have a problem of supply even now. The small and marginal producers in the remote areas will not be covered by private systems of procurement through 'middle men' on account of high transportation cost and perishable nature of milk which requires chilling and processing. If at all this is collected from distant areas it may be only during the lean season, but with all exploitative practices which are well-known and have been well-recorded in the past. Moreover, such incidental efforts will shake the confidence of the producer in their co-operative organisation. It is not certain whether the dairy industry after the new policy will be able to offer a price that is close to the actual cost of production. What is certain is that if there is a flood of commercially imported milk powder emanating from the pressures of the MNC from outside and the corporate giants from inside then, all incentives for production at home will reduce as the price of indigenous milk will get depressed by imports and on account of relatively cheaper imports even the desire to promote production and enhance procurement will get reduced resulting in negative growth of the Dairy Industry. Therefore, clearly the competition is unlikely to help producers of milk particularly the weaker sections among producers who constitute more than 7%.

(2) GROWTH OF OUTPUT : An open entry of MNC and DCS and their collaboration will certainly lead to rise in output but what type of output ? For whom ? At what cost ? It will lead to rise in the output of luxury products at the cost of 'liquid milk'. In fact, excess output of chocklates and cheese will lead to 'Milk Crisis". Though we have made considerable progress in the production and consumption of milk after the Operation Flood Project, we have yet to achieve the minimum milk consumption level even in urban areas. It has been reported that 15% of the consumers are not able to buy any grade of milk; about 30% can buy only half the per capita minimum recommended quantity, the next 48% of urban consumers can afford to buy just the per capita recommended commodities, only 7% of the urbanities have enough purchasing power to buy milk in desirable quantities.

Obviously, increased price of milk certainly discouraged the consumers. (a) The output of raw milk at the farm level will not be promoted as the benefits of competitive price will not be available to producers, (b) private traders as it is our experience are never interested in production promotion and (c) capacity of co-operatives to spend on production promotion will decline (d) the opening of the "Commercial Imports" in the name of import liberalization for export promotion will lead to disaster.

(3) PROMOTION OF INVESTMENT : Investment is a key to modernization. The Indian dairy sector was traditional and in-efficient due to the almost investment starving status of the dairy sector in the planning prior to 1970. However, under the Operation Flood projects, we invested about 1800 crores and developed modern infrastructure including modern dairy plants. Our current priority is to consolidate the efforts to achieve optimization of investment made in the modern infrastructure and plants. Therefore, the Operation Flood-III has rightly concentrated more on consolidation than on new investments. Moreover, as in the case of Operation Flood-III, further investments can be possible without having to depend on the EEC or any other source and hence, seems that any kind of case for open entry into the dairy industry on the grounds of investment will not sound convincing.

(4) TECHNOLOGY PROMOTION: Like 'investment', technology is also a misplaced issue. We do need to have a technology appropriate enough to promote production at minimum cost of production without replacing labourers. WE did not require 'technology' for the production of product manufacturing and processing prior to the Operation Flood, because India was highly dependent on every kind of dairy equipment and plants. However, thanks again to Operation Flood which encouraged Indian 'technology' and today we are able to have 'Indian technology' for all kinds of equipments and plants. The share of imported technology was as high as 70% in 1970 but today 130 units with 5 large public sector undertakings cover 85% of our needs as a result, the import content in our Operation Flood Projects II and III was not even 15% of our investment. In fact, some of the innovation like production of cheese out of the buffalo milk is India's original contribution to the world. Similarly, technological innovations of the India Cooperative Dairy Industry in the production of Indian milk products is indigenous. Further, investment in the western technology has also become irrelevant. Such technologies are highly capital intensive which by nature are characterised by low employment intensity and is even labour replacing in many cases. Thus, it is difficult to establish that there will be technological advantages in the adoption of the new policy.

(4) EXPORT PROMOTION : Export promotion on regular basis with significant volume through dairy products is a matter of very long term ambition and not the reality of today. Despite the reasonable stable growth rate of milk production, we have yet to generate adequate surplus on regular basis to produce products for exports. Any such effort could not be without the cut over the bare minimum level of per capita consumption of milk achieved by India after a long time. There is no reflection on the achievements of the Indian dairy co-operative sector, but it is quite well known that in terms of qualitative standards and competitiveness in comparison to the developed countries we have yet miles to go.

However, there is a point in view among knowledgeable circles that "the new policy will encourage more MNCs to promote exports of dairy products from India as it is cheaper to manufacture milk products in India and sell them elsewhere. Production costs in India are said to be much lower than in the European countries". Even if we accept this point of view from them the question to what extent India will benefit in terms of export earnings is difficult to understand after the recent change in "FERA" regulation and also after our experience with the MNCs. For example the NCAER study on "FERA" companies have shown that they are inward looking, capitalising on the 'phoren craze' in the country rather than helping to boost the countries foreign currency earnings.

Thus in this assessment of delicensing in dairy products, it is clear that many factors do not pass the test given the facts. In other words the logic behind the policy change do not seem to be operational in the case of "dairy sector".

IMPLICATIONS OF THE DELICENSING POLICY : The open entry of non-co-operatives will certainly lead to an environment where co-operatives will neither be able to work with viability nor make the replication of "Anand pattern dairy co-operatives" possible for extension to other parts of India. Thus, the degree of adverse impact of the policy will fall on the co-operatives depending upon the area and its relative competitive strength to survive against the tactics of the non-cooperative sector, the implications of which could be analysed as follows.

1. The co-operatives shall not able to compete with the corporate sector and the MNCs as they are not "EQUAL" because (a) By definition co-operatives are wedded to 'social good' and departure from this characterisation means that they are not any more co-operative but are corporate bodies. Therefore, it is less likely that it may leave people without liquid milk. (b) This function will not be possible by the dairy co-operatives without very high cost of procurement which they will be forced to pay to attract milk from the producers. (c) As the competitive strength of dairy co-operatives is not the same because perhaps except for Gujarat at many places, the dairy sector will face a 'crisis' of survival which will compel the co-operative dairies to divert milk for product manufacturing in order to drive adequate margins for survival. (d) In the area of product manufacturing, it is well know that co-operatives have to follow States Laws and also pay every kind of taxes etc. which are generally by-passed by the private sector. The recent story, of the Rs. 700 crores profits in black money earned by the (oil trader in Gujarat provides an example of the capability and credibility of the private sector and in business, competition strength is strength with total disregard to the sources of strength) (e) Finally, since the dairy co-operatives are committed to its producers, it can neither afford to play with other milk trading tacts like the traditional volumetric system of pricing nor can it get rid of certain production sector will have a very challenging time for survival and less opportunities for growth.

II. Secondly, the 'National Milk Grid System' which had ensured stable supply and price to metros and other cities of India will stop working as the system will require adequate 'surplus' to feed deficit areas and the new environment of open entry will not leave the co-operatives with adequate surplus to serve these areas. This will lead to "Milk Crisis" at least in the major cities of India in the summer. Unless we again allow 'Commercial Imports' to feed the plants to reconcile the crisis Howeve. In that case, our idea of earning export surplus will be straightway affected.

III. Thirdly, it was only after considerable efforts that the replication of Anand Pattern of dairy co-operatives started showing very hopeful trends. The States other than Gujarat were coming up rapidly. This process of replication of the dairy co-operatives will get stuck up now on account of the powerful trade tactics of the product manufacturers in the area of procurement as well as production. This is most likely to happen in the States like UP, Punjab, Delhi etc. where production is maximum but the co-operative structure is weakest mainly due to the private sector dealing with product manufacturer.

IV. One of the basic contribution of the co-operative sector in dairying was to achieve a break through in production and productivity, despite the fact that one may not agree to the exact correlation between the output production and the Operation Flood project areas. However, the fact remains that prior to Operation Flood in 1970 the growth rate was just 0.7% per annum. This production however indicates a clear rise in the production per animal. The capacity for financing production enhancement programmes will decline with the deterioration of the viability status of the district unions. The process of production promotion at the farm level will be stuck because the matching efforts to keep the level will not be available from the non-cooperative units. This has been a story in the past and will become the story of the present. For example, a case of the private sector intervention in the promotion of Coca production in Kerala depicts the true face of MNCs.

V. Finally, the most important implication because of the change in the policy will be the adverse impact on the rural poor. It is well known that Indian dairying is an economic pursuit of the small marginal farmers and landless producers. It was through the co-operatives that they were able to realise the potential of the dairying as the real source of income and employment. Under the new situation growing privatization will not provide real scope of development and declining co-operatives strength will be found inadequate and ineffective to promote their upliftment through dairying.

SOME CORRECTIVES TO COMBAT CRISIS

The basic course of correcting the crisis is to withdraw the policy of delicensing and to provide regulation coverage to co-operatives atleast for sometime. However, in view of the Prime Minister's almost constant utterances of promise of "NO REVERSAL" to "ECONOMIC REFORMS", the chances of such change will be unlikely in the near future and in fact, the request of the Chief Minister of Gujarat is already reported to have been rejected by the Prime Minister. Under this situation can we create an environment to balance the position of the co-operative sector and reconcile the conflicting interaction of the producers, consumers and also the State?.

1. First of all let us accept the words of wisdom of the President

of the International Dairy Federation (IDF) Mr.Dykstra who said, "By heart and conviction, I am much in favour of liberalization or delicensing as you call it... however, we must recognise that for the agricultural sector, such liberalization has never occurred in any country even in the most dairy advanced countries with full scale liberalization without protect of the interest of consumers, farmers and society at large. Some institutional mechanisms do exist i.e. Marketing Boards in U.K. should we not think in terms of developing an institutional mechanism to regulate the private sector in India ?

II. Secondly, instead of relieving the co-operative dairy sector from liquid supply obligations, the corporates and MNC and their collaboratives must also be asked to supply some proportion of their procurement for liquid milk supply to the consumers of their towns and metro cities. No dairy product manufacturing unit should be allowed to enter into product manufacturing alone.

III. The regulating Act of the co-operative sector should be equipped with new provisions. An outdated current co-operative act hampering autonomy of the co-operatives and increasing the bureaucratic interferences must be changed as the recent note of the Ministry of Agriculture has shown how the viability of the dairy is affected by this Act. Fortunately, the Planning Commission is already ready with the Model Act of Cooperatives which must be immediately accepted to strengthen the viability of the dairy cooperatives.

IV. Two-axis pricing method whereby the procurement price is fixed by qualitative measures of FAT and SNF must be compulsory and particularly the non-cooperative sectors must not be allowed to follow the traditional volume based method of pricing whose exploitative characteristics have been proved beyond doubt.

Two axis pricing will provide for great protection to farmers.

V. The distinction of the use of milk for liquid consumption and product raw materials should be recognised and 'Milk' as the raw material should be taxed to reduce the profit margin of products which will help in keeping the balance in the utilisation pattern and also help in controlling supply shortage due to over divergence of milk in product lines. How long it is feasible is a matter of the details to be worked out.

VI. Institutional mechanism should be provided to consumers by establishing a pricing board on the lines of the U.K. Milk Marketing Board and the producers should be protected by accepting their long due demand by including 'Milk' as a commodity for consideration for cost of production by the cost and Agricultural Price Commission which regularly does such exercise in case of agricultural crops. If the product manufacturing units are likely to enjoy fat profits why should not the milk producer be priced according to the cost of production ?.

VII. Finally, all the products manufacturing units and co-operatives also should be compulsorily asked to spare some percentage of their profits for the supply of technical inputs to producers so that adequate production and productivity could be achieved to balance the supply-demand gaps arising out of the new policy.

Thus, the new policy of delicensing will not be beneficial to consumers, producers and even to the Government. Some suggested measured may perhaps help to reconcile the interests of all the classes and also the State to some extent.

□□□

6

Dairy Product Export Performance and Policy in Post-Liberlisation Period

Introduction :

This paper is an attempt to examine the possibility of increase in rural development opportunities through promotion of dairy product exports on account of expected favourable implication of the GATT/WTO. The design of the paper is composed of the following sections.

First section depicts dairy development scenario along with some recent changes in the dairy policy of India.

Second section examines export status of the Industry and assesses the export potential.

Third section analyses the GATT implications for the Indian dairy Industry.

Fourth and final section suggests the basic policy action to get the best possible advantages of whatever opportunities are likely to be available to Indian dairy sector. The paper ends with an appropriate concluding note.

I

INDIAN DAIRY INDUSTRY SCENARIO AND RECENT DAIRY POLICY CHANGES

Dairy Industry of India prior to 1970-71 was characterised by almost stagnant marginal rise in Milk production growth i.e. 0.7% per annum resulting in to falling per capita availability of 133 grams of milk to 100 grams in 1971.[1] Consequently, import dependence of India was as much as 60% in total milk throughout in India.[2] Three phases of the Operation flood projects transformed this traditional dairy sector in the modern dairy industry. How it could be done is now very well known and well documented elsewhere.[3] We may very briefly record the major achievements which made this transformation a reality.[4]

(A) National Accounts Statistics records that milk is now India's largest crop after rice as total value of paddy was Rs. 12808 crores and milk group had value of output of Rs. 10573 crores. While third largest crop had recorded its value of output at much lower level of Rs. 7770 crores.

(B) This is the outcome of the very high rate of growth around 5.68% not only higher than population but also agricultural and Indian economic growth rate recorded during 1950-1990. Obviously, this led reversal of per capita availability from 112 in 1972 to 198 grams and in Metro cities as high as 313 grams.

(C) This production breakthrough is real and sustainable as it was consequence of not only size but also improvement of yields per animals from say 600 kg. in 1972 to 1476 kg per year in 1990 i.e. 146% rise. Apart from technical inputs even rise in crossed animals increased at the 9% per annum also contributed the yield promotion.

(D) The gradual transition of Indian dairying from subsistence to modernization has reflected in the enlarged size of the marketable surplus and increased share of the organized sector in the market. It is estimated that about 75% of milk is consumed by household sector. 10% is marketed by co-operatives, 2% milk is handled by big private units and government schemes. 275 liquid dairy plants and 83 dairy product factories are working at reasonable level of capacity utilization of 62.5% and 79.2% respectively. The demand for milk is growing at reasonable scale due to favourable income elasticity but the pattern of milk utilization has broadly remained dominating in the favour of liquid milk as estimated consumption of milk as liquid milk is 50% followed by 28% ghee while butter and curd accounting for 6.5% and 7.0% as a result the product market served by organised dairy sector accounts for not more than 1% to 1.5% of the total milk production.

(E) Indian dairy scenario will be incomplete without taking into the consideration of two major distinguished phenomenon. One cannot afford to ignore the fact that as R.K. Patel has put "milk production is predominantly the domain of small holders in a mixed farming".[5] 70%-75% of small households possess milch cattles as per NCA estimates also endorsed by NDDB numeration survey of 1984 in 20386 dairy co-operative villages under operation of flood milk sheds). Relevant Indian dairy policy should have to protect and promote the interest of this class as the central objective of the policy.

Another great validated fact is that impact of efficient producers oriented Anand pattern dairy co-operatives have help making dairying a very viable enterprise alongwith a great support of technical inputs leading to rise in yields and reduction in cost of production. As a result "A great economic and social transformation has taken place which could have not been done by any large scale development programme in India. One could find many evidences indicating such transformation in various studies.[6] Moreover, it may be suggested that recently Tushaar Shah has very effectively explained how the process of impact of dairy co-operatives operates on productivity and technical change involving tiny milk produced through dairy co-operatives.[7] We may, however, note that NCAER estimates that dairying generates 14.8 crores of person days of employment in India and also pumps as such 30,000 crores per day cash income. Thus, Dairy sector is so much dominated by the small and marginal holders that the National Commission on Agriculture had rightly recommended the "Milk Production Promotion through small and marginal milk producers."[8]

Recent Fundamental Dairy Policy Departures

Against the national policy of developing dairying through small and marginal producers by promoting Anand pattern; as a part of "Liberalization" dairy industry is also delicensed in July 1991. We have discussed its logic and reality in our earlier paper.[9] However, this great policy departure by way of exempting dairy processing Industry from Industrial Development and Regulation Act of 1951 Which formerly prevented the entry of private non-cooperative dairy operators in the industry.

Another policy change in the favour of multi-nationals was that they were allowed to raise their equity holding upto 51%; gave an extra-ordinary strength to all parent companies of NNC in India to raise their equity holding. Glaxo and Nestle who had not more than 40% increased their equity holding to 51% and then, strengthen their base and interest.

The very first consequences of opening up the Industries were found in terms of numerous plants come into existence without adequate desirable hygienic standards, most of them meant for product manufacturing only. Moreover, co-operatives strongly opposed the policy on the ground that private units are taking ready crop in terms of Milk supply and yields were co-operatives had spent decades of its monetary investment and manpower. Infact, Co-operatives felt truly threatened. Out of 689 applications, 95% of there were for product manufacturer. Moreover, some states attracted the investment by incentives As a result, capacity exceeded the potential production. In fact, this concern for the excess capacity and excesses in misconduct and mal practices prompted call for some kind of regulatory measures.

In July, 1992, Government of India promulgated Milk and Milk Product Order (MMPO) to control on entry and operation of a game play in opened dairy sector.[10] Though critics regarded this as reintroduction of regulation by backdoor. However, the provisions of getting legislation for handling more than 10,000 ltrs. per day, state permission and control government permission after some limit; and compulsory liquid milk sale along with development of their own milkshed etc. were the provisions helpful to get regulated and balanced privatization, however, the MMPO is very heavily criticized on many ground and as a result, in May, 1993, MMPO so modified that co-operative sector found it "Toothless" because under the diluted MMPO, registration requirement is abolished upto 75000 11 pd or 3750 Mt solids and renewal term period also increased from 3 to 5 years and advisory body MMPO also reconstituted to regulate the strength of co-operatives. It is feared by NDDB that the capacity creation has already exceeded the quantum of marketable surplus available.[11]

Thus, recent policy changes are towards liberalization at the cost of co-operatives and producers. The marginalization of the weaker co-operatives particularly in non-Gujarat area hasdeveloped terrible institutional vacuum on account of this, for the small producers would extremely constrainous to promote the interest of tiny producers of the dairy sector. Under this state of affairs, we may examine the Export status and export potential of the sector.

II

EXPORT PERFORMANCE AND EXPORT POTENTIAL OF THE DAIRY SECTOR OF INDIA

International trade in agricultural sector is almost constantly declining phenomena with marginal fluctuations within the range of 22.8% in 1951 to 20.08% in 1980. Then after, it declined with quantum down to 6.5% of total exports. Similarly, imports was as high as 28% in 1951 of total imports which continuously declined falling at the bottom 6.8% in 1994-95. The interesting explanations regarding this phenomena is well observed in the analysis of structure and composition of exports and imports elsewhere.[12]

Within this declining significance of agriculture in international trade; obviously milk and milk products are so insignificant that until recently, milk and milk products were clubbed together in livestock exports. For the same reason of its insignificance; the 8th plan did not spell out any specific targets of exports of milk and products. Livestock product accounts for hardly 4% of agricultural exports; within the livestock it is leather products with 50% share constitutes a major product and buffalo meat with very high 8% rate of growth of exports is regarded to be as more significant product of export than milk and milk products which constitute not only largest share in livestock output but even second most significant crop just next to rice in terms of value of output recorded In 1992-93, with just as 3 million worth export constituting just 0.1 of total agricultural exports shows export significance of dairy industry.[13]

Recent liberalization policy is extended to export promotion by special incentives. Dairy Industry has got its due share. Apart from decanalization of exports and special promotative role played by APEDA. It removed by "Pre-shipment Inspection etc. export Incentives are redesigned but export licensing and exports ceilings remained with APEDA. Since liberalization of dairy trade what has happened to dairy product exports in term of quantum, values, structure of products and export markets spots etc. would be quite interesting and useful to observe which could become possible due to availability of data made public by APEDA.[14] Based on this data we may draw the following interesting and useful observations.

1. Between 1990-93 dairy exports continued to only 3 to 5 products namely milk cream and other sweet material, butter and ghee, cheese and curd, natural products and accidified milk and cream.
2. Though quantum of export is not very significant in terms total exports has grown by 7 times in post liberalized period.
3. However, the observed rate of growth was quite erratic.
4. Major exportable products with substantial quantum are only two i.e. Butter and cheese and milk and sweet cream. In 1993 cheese with 23% share substantially reduced the share of butter and ghee.
5. The Indian dairy exports have mainly attracted Russia and neighbouring countries and some Gulf countries each of these country has shown interest in particular dairy product in particular time.

   Sometimes some marginal quantum of exports of some products are made in particular countries may for some specific political objectives. Exports of natural milk to Somaliya illustrates the point.
6. Our major buyers of milk cream are Bangladesh, Russia and U.A.R.
7. Our most of the butter and ghee are exported to United Arab Amirat and Bhutan. It may be noted that these products constitute 80 to 90% of dairy products.
8. We may finally conclude that our main export markets are Russia, U.A.R., Bahrain and Bangladesh.

Thus, specific product and countrywise concentration even in small quantum of irregular and seasonal exports become obvious by way of the analysis of the exports in 1991-93. The available reports of 1994-95 shows clearly that exports of dairy products is simply conditioned by our surplus which is uncertain and erratic. A recent report states that milk product exports which have been increasing in recent years crossed 13000 tonnes milk products in 1994-95.[15] However, Government has erected a drastic cut in the export ceiling for milk powder to 10,000 tonnes from 25,000 tones in 1994-95 fearing spurt in the prices before Loksabha elections the ceiling for other milk products such as ghee and butter has also been reduced to 1,000 tonnes against 7,500 tonnes. The Director General of Foreign Trade has already issued a notification that only 10,000 tonnes of milk powder and 1,000 tonnes of pure milk, ghee and butter can be exported this year (1995). This despite the fact that exports in 1994 was less than ceilings fixed up. Under this situation, exports has found to be playing negative role of raising prices of milk. NDDB for example in its Report of 1994-95 it has mentioned that "the reported export of a sizable quantity of milk powder in 1994-95 by private sector at a price lower than domestic market price, created a further local shortfall.[16] In fact some estimates suggest that powder was exported roughly equal the imports likely to be necessary to ensure milk supply".[17] The evidence is substantial that there will be sizable shortage of milk particularly on account of excess capacity of 937 lakh litre per day against estimated marketable surplus of 500 lakh litre per day.

Thus, export scenario of India clearly shows a modest initiation of exports of selected seasonal products concentrated in some developing countries with erratic trend.

Imports of Dairy Products in India

Before we discuss the dairy export potential of India in near future, we may take note of some facts relating to imports.

The story of development of Indian dairy Industry is well expressed in its progressive reduction of imports or for that matter its achievement in import substitution as well if we include related imports like that of dairy equipments etc.

It may be stated that important component of India's MIlk throughout year as high as 74.6% in 1951/52. All these imported Skim Milk Powder and Imported Butter oil was commercially imported. The range of imports was as much 30% even in 1970s. Thanks to OF and NDDB that Commercial imports were totally ceased from 1975-76 and gifted imports also went down continuously, from 10.1 to 2.3% without reduction in consumption as throughut from indigenous sources went on increasing due to promotion of production. For example in 1989-90 the total through put was 749.9 lakh litre per day and yet the import comprot was just 2.3% of gifted materials by all means. This is an achievement.

However, on account of first export of small quantity of 3500 MT of Milk powder which was surplus and was having possibility of quality deterioration. The EEC, dishonoured the commitment to provide 75000 MT of SMP and 25000 MT of Butter oil. The balance quantity of 15000 MT of Skim Milk and 11500 MT Butter was straight away denied by the EEC. In the following way :

"India has been successfull net exported of Milk during the course of last year which demonstrates the country's capacity to play an active role in the international dairy trade. It is felt ,therefore, the strategies of this magnitude should be balanced "Primarily through Commercial Imports". The *fact* that deficit has been reported in a particular year does not argue for the automatic allocation for food aid."

This EEC failure to honour the protocol of 1988, resulted into acute shortage of SNF and Fat and compelled NDDB for resorting to "Substantial Commercial Imports" of Milk powder and butter oil. NDDB believe that this stand taken by the EEC is because of their concern that India is achieving self-sufficiency in milk and has entered the export market.[18]

As the information with Ministry of Agriculture during 1994-95 a total of 4305 metric tonnes of butter oil was imported on a commercial basis at price ranging between $ 1600 1865 per tonne. The higher price were the result of a world-wide shortage of milk fat.[19]

Import of butter oil had, no doubt, helped increase the supply of liquid milk by facilitating recombination. The average selling price during the period was Rs. 84333 per tonne in the domestic market. In the pervious year i.e. 1993 there was no commercial imports of butter oil while in 1992-93 about 402 tonnes was imported.

In February, 1995, NDDB anticipated shortage of milk, it contracted for import of 6500 MT. milk powder in February, 1995. This was done to avoid disruption of liquid milk supply in metro-cities and other major towns.[20]

Thus, we had achieved a self-reliance by complete import substitution since 1975 thanks to operation flood projects but now slowly tables are turning up partly because of policy reversal and partly due to bad monsoon caused shortage alongwith withdrawal of EEC aid assistance at this critical moment, however; we do hope that India will soon get recovered in near future.

Before we pass on to assess the export potential of India in dairy products we may take note of relatively lesser known but really significant achievement in the area of import substitution of dairy equipments.

Import Substitution in Dairy Equipments

Another remarkable achievements of India is "Self-reliance" in dairy equipments and dairy plant construction know-how[21] under the operation flood attempted to develop modern dairy sector and progressive reduction on imported equipment is very fast achieved by 160 units working now, earlier only 15 units were existing out of 160-109 units are in small scale sector, 43 medium and large and 8 are in public sector). Progressive reduction of imported equipment could be observed by the fact that between 1972-73 to 1988-89 import dependents in dairy equipments have been reduced from as much as 17.14% to 1.66% despite the fact that demand for equipment which was the value of 40 million has expanded to Rs. 2000 million due to development of the dairy sector in India.

III

EXPORT POTENTIAL FOR INDIAN DAIRY SECTOR

After reviewing export scenario of Indian Diary Industry, we may take up the question of export potential of dairy sector in near future. We may judge the export potential of dairy product in the light of the position of India with reference to export related issues such as exportable surplus, international competitiveness of Indian Milk and Products and the potential demand of Indian dairy products in International trade.

(a) Exportable Surplus

(There are many studies relating to demand are supply related milk in 2000. A.D). (One authentic study of Misra estimates a economic demand and nutritional demand based on expenditure elasticity of milk at 1.18 and nutritional requirement at 250 gram per day respectively). The estimated economic demand works out to be as 97.70 MT in 2000 A.D. from the level of 35.81 MT in 1982, and Nutritional demand works out to be 91.28 in 2000 A.D. from the level of 35.81 MT in 1988. Against that at the 4% rate of growth of milk production per annum there will be hardly any possibility of meeting even demand what to talk about surplus ?[22] This is because at the rate of 4% per annum, projected production of milk in India will be 78.14 million tonnes at 4.5%. It may be 81.59 and at most liberal rate of 5% it will be at best 85. In leaving a gap of 19 million, 16 million and 12 million. If we consider the economic demand which is really effective demand. The significance of the gap between projected demand-supply could be best appreciated if we relate our import needs in early 1960's which was around 20 million on an average. Thus, chances of new surplus are less likely to occur, unless we take advantage of a seasonality and milk cycle to mop surplus after developing required buffer stock to compensate our needs in lean period or slack years. However, incidentally, Kirit Parikh's, study estimates optimistically of 3.39 milk equal 10^6 tonnes of dairy products if India achieves 4% rate of growth in agriculture.[23]

(b) International Competitiveness

Indian milk price in comparison of several developed countries is regarded to be as low due to several reasons including much less component of marketed input in output are able to maintain the cost of production at lower level which is indicated as follow :

Country	Milk Price $ US per Kg.	Country	Milk Price per Kg.
Switzerland	12.75	U.S.A.	4.93
Japan	9.52	India	4.17
Republic of Korea	8.67	Australia	3.74
EEC	5.61	New Zealand	2.21

Source : R.P. Aneja, "Pricing Policies and Marketing of Milk" in Dairy India, 1992.

Since 'Milk' is the major component of the dairy product the cost advantage in production seemed to be obvious. However, there is another point of view also. It is estimated that the margin between producer's price and whole price are much larger in case of India indicating marketing and processing inefficiency. For similar reasons, the dairy products are costlier for example butter price was quoted $ 3743 per MT in India, $ 2251 in USA and world average price was $ 1498![24] Therefore, just by having relatively less cost of production of milk it would not be easy to arrive at favourable competitiveness. Further, privatization may provide still better price to producers in that case we may loose this advantage in future, however, there are chances to close the gap between India and international market in case of some products provided marketing and processing efficiency is further achieved.

The market for Indian dairy products certainly could not be possible in developed countries on account of that they are generally having "surplus" In fact, a study for the FAO shows that by 2000 A.D. the developed countries would have 17.5 million MT of surplus milk per year in contrast to deficit of 25 million tonnes in the developingh in the dairy sector. It is estimated that annual import by developing[25] countries is expected to be 43 million tonnes products, which also represent a major challenge along with competition from surplus generating marketing seeking developed countries, however, with growth, there will be great demand for milk and products in developing countries due to high income elasticity for milk and products against that developed world will find another contracting influence over dairy product demand is that "In the western countries the value of milk fat has been put in doubt on ground of health consideration". This anti-dairy product image is threatening to come up under the pressure of aggressive advertising[26]. As we have seen in our export scenario of India, India has reasonable opportunity of exporting dairy products in the Gulf countries, neighbouring Asian Countries, and of course in all Sark countries due to SARC, similar income distribution pattern in favour of middle and upper middle class along with great cultural, traditional and quality and test similarities India stands great chance in these countries for the promotion of dairy products.

The obvious response to cost advantage are comparative advantage in favour of India in case of dairy product could be negative, however, this may not be true in case of "Indigenous traditional dairy products" which could be popularised as exportable among many developing countries and even among developed countries provided we could developed sound technology to produce some of our products out of "cow milk", however, the advantage in indigenous products is possible to draw on account of much less major raw material cost. For an example following shows the raw material cost of the products as percentage of sales prices.

Sr. No.	Products	Percentage
1.	Market milk (bulk vended)	90%
2.	Market milk (packed)	80%
3.	Milk powder	70%
4.	Butter	65%
5.	Pandda/Burfi	35%
6.	Rasgulla	39%
7.	Sandesh	39%
8.	Gulab Jamun	34%
9.	Shrikhand	29%

Source :- **R.P.** Aneja, Ibid, 1992.

The operational experience of developing indigenous product with improved technology and high standards and quality and hygiene has proved much successful in deriving greater returns out of the business is an established experience set by "Sugam" dairy plant of

Baroda[27]. Export potential of all these products are very high in developing countries because of high income elasticity of demand for milk and products ranging from 0.5%, 1.2 in developing countries along with fast expanding milk demand on account of new milk consuming countries such as Western Africa, south east Asia, where there is a little tradition of consumption of milk and products are also offering demand. Thus, new and larger markets with lesser direct challenges of Western countries likely to be available to India atleast for the short period . Until developed countries also evolve some technology break through in developing indigenous dairy products of India which is most likely to happen due to "open door policy" followed by India since 1991.

Thus, our analysis of dairy sector from the point of export possibility today and tomorrow tells us clearly that:

i) We are likely to secure a thin deficit of milk instead of any regular "exportable Surplus" without causing any internal disadjustment.
ii) Price differential between Domestic International price also is less likely to be attractive for exports atleast in the near future.
iii) While there is no scope for the promotion of exports in developed countries and western products; we may have reasonable opportunity of exports in developing countries particularly in exports of indigenous products.

WTO Implications for Dairy Sector

It may be noted that GATT is an international trade arrangement and aiming at "Free Trade" implying liberal exports and imports.

There are six key features of General Agreement on Tariff and Trade[28]. Which could be relevant for Indian dairy sector.

1) tariffication of all non-tariff barriers.
2) average reduction of restrictive tariffs by 24% over 10 years.
3) Reduction of export subsidies by 24% over 10 years from a base period of 1986-90. Reduction of aggregate subsidies by 13.3% over 6 years.
4) minimum access of imports equivalent to 3% of domestic conservation with a 1986-88 based period raising to 5% over 6% years. If the access is greater than 5% this must be maintained.
5) Domestic support (subsidies) of less than 10% and "green box" polices are exempted, i.e. general service policies, direct payment to producers i.e. social safety nets and development policies and investment subsides and agriculture inputs, subsides primarily to low income producers. Other forms of domestic support classified in "ABER BOX" policies are subject to reduction.

6) Restriction of price of basic imports are disallowed unless, granted by the committees on Balance of payments on the ground that increased imports will be detrimental to its balance of payment situation. (It is further allowed to its quantitative restriction. If priced based movement can not arrest the sharp deterioration in the external payments position. India, however, has already announced the time table for its removal).

However, on account of specific conditions of India related to adverse balance of payment, less than prescribed limit of aggregate subsidises and development subsidizes supports below 10% of agricultural production etc; WTO implications for agricultural sector with reference to existing income support and export subsidy will be no more, so far as dairy sector has concerned India provides no kind of subsidy to producers at any level. It will not have direct adverse impact, however, as we shall see later there are some other adverse impact emerging out of trade liberalization and consequent Import-export decanalisation.

Since new strength has acquired by dairy Industry of India in last two and half decade as a result, in the long term dairy industry expected to emerge as foreign exchange earning Industry of India.[29] The logic of this expectation of course is based on number of assumptions related to policy changes and behaviour of major countries like USA and EEC etc.

With establishment of WTO world trade in dairy products likely to be more open. It would also make the field more competitive. In 1992, the world traded milk worth US $ 1 billion, US $ 9 billion in butter, and US $ 3.5 billion in cheese.

Milk production is heavily subsided in Europe and the pacific rich countries. European common wealth producers provide 63% subsidiary to dairy products, 67% in the USA and 60% in European and pacific rich countries. Infact, except New Zealand all countries in the world subsided dairy products. The acceptance of GATT will benefit the Indian dairy sector as with GATT/WTO the subsidy for the milk producers in Europe will go, leading to steep rise in milk price in their markets. While Indian prices are already competitive will become increasingly competitive and attractive to the world due to this probable impact of GATT in developed dairy countries.

In some markets such as the USA, support prices were too high to permit competition in the International trade, forcing the government to assist export programme under these circumstances, New Zealand and India, both giving no special treatment for dairy farmer could offer milk products at competitive prices.

However, the GATT benefits to India are doubted on many grounds. Overall GATT benefits to world will be an increase of 235 $ billion annually and by 2002 it will be of the order of $ 735 billion, however, largest part of benefits will be mopped by developed countries whose share in the world trade is far higher than developing countries.

Moreover, there are doubts that developed countries will not honestly implement the GATT agreements at all. The direct indirect evidences could be stated at follow:-

1. Recently the WTO has been interested with non-flow related issues like **social-clause** and environment clause.30 Earlier issues like child labours, labour standard, etc. were dealt with by ILO, now it is taken over by WTO, in the name of effective mechanism by linking them with trade. In fact, three or four social issues made relevant in WTO operation.

   I ensuring environment friendly trade

   II evolving labour standards pertaining to child labour.

   III ensuring employment and

   IV preserving community and culture.

   These are enough grounds to create artificial barrier to trade in the world by the developed countries. Dairy Industry also could be victimized on such grounds.

2. There are evidence to believe that developed countries which are so sensitive to the entry of developing countries in international trade that as soon as they find; India becoming second largest country in milk production and even exporting country. They would strongly react. It is well known that how, EEC discontinued its aid obligation to India on the ground of India's modest entry in international trade[31] i.e. India in the whole region of Asia and Africa only country to challenge the developed countries.

3. On the basis of EEC reaction of India's export efforts, it is quite logical when Kurien expresses his legitimate fear that GATT will not be honesty followed at all.[32] ("Unhealthy" **trading habits of subsiding and dumping** that were permitted by this same GATT!) Kurien believes that there are many ways to avoid provision of withdrawing **subsidies on dairy products** to prevent India and other countries from capturing their export market. (There are indications that developed countries may provide a large subsidies on **processing**. In the place of production and thus status quo is possible without loss to producers and markets. GATT restricts subsidy or **Production Only**).

4. The GATT provides for 3% of Imports of the domestic market of all commodities which would mean compulsory imports to that

export in dairy sector. The quantum of such Imports cannot go without depressing effect on domestic market however, due to exemption available on the ground of adverse balance of payment. There is no obligation of compulsory dairy products Imports in the immediate future. However, chances of dumping due to decanalization could not be avoided.

5. We should also note that even after sincere implimentation of GATT of India hope to be a major player particularly in south east Asia if not in Europe it needs to find new pastures/marketing besides attempting diversification and production of internationally standard quality production like cheese better milk power, Ice cream and chocolates unless India develops right mix of dairy products it can not expect to gain hold in the world dairy market. Dr. Kurien has rightly said "neither is the world waiting for our milk and products nor are the advanced dairying nations going to sit idle by which we grabe their market."[33]

6. Developed countries in dairy trade create great obstacles in milk product exports; they could do so is well illustrated by NDDB export[34] who reported that India had raised strong objection to "Technical barriers to trade (TBT) at the international dairy federation. European union and the USA have already implemented some comprehensive technical regulations for various goods including dairy products recently. The prejudice of western countries against "buffalo milk" gives an example of technical barriers to trade in the place a economic barriers of pre WTO period. A codex committee to review and revise the existing standard in view of the changes anticipated by Implications of the GATT agreement restricted its scope to cow milk only and buffalo milk was totally ignored in **definition of milk.** However, lot of efforts were required to be made by IDF members including India to revert to the definition of milk originally included in the codex standard of milk and products and thus, attempt was made to exclude buffalo milk and its products from the traded goods in international trade particularly in case of all those developed countries who strongly follows the norms of International codex of quality.

7. A recent report of UNCTD on trade and development also may be quoted to provide an evidence of tendency of developed nation to discourage directly nations in taking advantage of Globalization and liberalization[35] . The "protectionist" practices followed by the developed countries are likely to be continued even after GATT agreement of 1995. The Indian experience regarding woolen

garments quota dispute between India and USA for which we had to go to WTO illustrates the point. In fact, as the UNCATAD reports points out the "the fallacy of the developed countries thinking that their unemployment is solely due to cheap imports from the developing countries".[36]

8. In is interesting to note that 22 members of the US Congress has openly brought out "GATT as **bad** for American dairy Industry". They have felt that the entry of impact of dairy products due GATT will certainly displace the domestic products. Therefore, they have asked for Export incentives and tariffs etc.
9. Finally, there is a great threat of dumping of imports due to decanalization depressing Indian production.

Thus, behaviour of developed countries in the international trade discouraging developing countries to promote their exports uncertainly and inherent incapability of WTO to regulate powerful nations; dominating influence of USA and unorganised developing countries on international trade front are less likely to provide benefits of GATT, on the contrary it will be most likely that developing countries may be used for dumping their surplus in poor countries affecting adversely to poor economies.

In the light of our earlier analysis of export performance and export potential, GATT seems to be much less positively relevant to Indian dairy industry atleast until near future.

IV

POLICY SUGGESTIONS FOR PROMOTION OF DAIRY PRODUCT EXPORTS WITH RURAL DEVELOPMENT BENEFITS

On the basis of our critical analysis of the present dairy sector, policy changes, export status and export potential and uncertain Implication of GATT/WTO under the present circumstances, what could be suggested to derive maximum benefits in favour or of rural poor by developing dairy Industry and to canalize the benefits of liberalization at home and globalization abroad for the rural economy and its poor people.

1. Indian dairying is reported to doing a business worth Rs. 52000 crores annually and it is expected to rise further to around Rs. 85000 crores by the turn of the century, however this economic progress of the sector fundamentally lies in coop development and protection or interest of producer and consumers and hence, every efforts should be made to continue Anand pattern cooperatives to be replicated in non-Gujarat area by strengthening state federations which could be impossible without Enacation of

inter-state co-operative act by central government in the line of model cooperative Act of planning commission.[37]

2. Similarly, even stronger cooperatives also will have to compete with non-coops unless they are put up on equal footing in terms of rights and duties with corporate sector. We have discussed at length elsewhere details of the needed changes in the co-operative legislations in the states of India.[38]
3. Decanalization of exports Import policy not only must be reviewed but also NDDB should be asked to under take this work in the place of APEDA as NDDB since 1987 have not only better experience but greater specialized Insight and expertise of Interpreting export market. A permission of exports 10000 MT of milk powder in 1994 without milk price given by APEDA shows how efficiently it works as how blind it is to mope up export earnings.
4. It MMPO is really suppose to operate to regulate the entry of private units then, toothless MMPO may prove dangerous in the long run often due its inability to regulate private units for example it is reported that even simple compulsory obligation of providing regular business operation data is not submitted to the controller of MMPO by private units. Many other illustration could be provided to prove helplessness of MMPO to regulate behavior of private units.
5. To promote exports, we must improve our production and yields to generate adequate surplus on regular basis to participate in export trade. NDDB has rightly stated that productivity is the future of Indian dairy Industry. Apart from intensification of traditional technical inputs distribution of cattlefeed, fodderseeds, and artificial insemination which has been done very well by 25360 dairy copps in India. We will have to make efforts to popularise cross breeding and application of newer methods of improving our animals stock by way embryo transfer etc.[39] (Our exports will be much affected adversely if cooperatives are not encouraged and privates are not compelled to invest in 2055 etc.)
6. We must also undertake specific measures to promote exports of dairy products, some of the measures be as follow :
   - I Export of animal feed like oil cakes must be stopped. A 10% export of oil crops lead to reduction of 3 million ton of indigenous output of milk. Feed scarcity facing country can not afford to do so.
   - II We must concentrate on export on result basis with specific export strategy. As stated earlier, we may develop a strategy

of concentration on some countries of Asia, Africa and Gulf countries to establish regular market with stability of standard supply. This could be done by promoting indigenous products with improved technology of its production

III The "cow" milk can be quite useful to promote exports of milk and milk products. In fact private sector may be encouraged to promote cow milk in those area which have not been covered by cooperative dairy sector.

IV The high standards of quality must be maintained. Unfortunately business practice followed by private and small dairy operators have led much below standard of quality including hygienic security. There are reports where the illegal practices of torturing cattle for extra milk by use of oxytocin, use of adding preservatives like soda to preserve milk by private and milk units engaged in preparing any selling "synthetic milk" comprising such components as water, refined oil, caustic soda, urea, common salt and sugar have been unearthed by police even and relatively known private dairy enterprise is made to pay Rs. 2000 compensation for selling adultered ghee as per the report in the pioneer of Delhi unless the laws related to quality resolution are made striker including making milk adulteration as non-bailable offense and the monetary penalty also should be increased. How can we think of export particularly through private sector when even ordinary rules of the business are not followed.

V Finally, there is one another area of export as India's dairy expertise has started becoming much in demand due to saga of success of Anand pattern and its replication in India. It has been appreciated by China, Pakistan, Cylone, African countries like Nigeria. On request FAO has also made its clear stand and support to promote Indian model in Asia and African countries. Infact, if well design plan of " North-south collaboration to build up dairy Industries" becomes a reality then, India will have a great opportunities of exporting skills and equipments.[40]

Concluding Note

In the light of of the current dairy industry scenario and recent dairy policy changes examined by us, WTO/GATT implications are very unclear. Also because, there are legitimate doubts regarding its real implementation by developed countries. However, in view of the present dairy policy, we are likely to get disadvantages of "imports" via dumping and deliberate imports by private sector while export of milk products are uncertain. In order to get best of all possible opportunity with percolative effects for rural development via tiny milk producer's co-operatives, we must take specific actions to adjust and restructure our industrial and trade policies.

REFERENCES

1. S.N. Misra, **Cattle Development and Dairying Policy Requirements Analysis**, Annual No. Yojana, Vol. XVII, No.1, January 1993.
2. S.N. Misra, **Ibid.**
3. See for example :-
   a) Dilip Shah, "Co-operative Dairy Development in India" Economics Times, Part I and II, January, 1991.
   b) Dilip Shah, "Replicability, Dependency and Equity Issues in the Operation Flood Projects", Journal of Political Economy, April-May, 1992.
   c) Dilip Shah, An Economic Analysis of Co-operative Dairy Industry in India, 1990, pp.25.
4. See for the data, R.R. Gupta (ed) Dairy India, 1992.
5. R.K. Patel, Present Status and Promise of Dairying in India Presidential Address at the Salde Conference of Indian Society of Agricultural Economics, 1992.
6. See in Example
   a) National Council for Applied Economic Research (NCAER) An Evaluation of Operation Flood Projects : 1990-91.
   b) Slde Roger and George Mergoes, "Dairy Development and Milk Co-operatives", World Bank, paper No.5, 1988.
   c) G. Parthasarathy, "White Revolution, Dairy Co-operatives and Weaker Sections", Economics and Political Weekly, 1981.
7. Tushar Shah, "Institutional Structures for Dairy Development, India's Post Independence Experience", Indian Dairy Man, January 1995.
8. Ministry of Agriculture India, National Commission Agriculture Interim Report of NCA on "Milk Production" through small and Marginal Farmers, 1971.
9. Dilip Shah, "Delicensing Dairy Product Manufacturing : Logic and Reality", Financial Express, May 8 and 10, 1992.
10. Ministry of Agriculture, "Milk and Milk Product Order", June, 1992.
11. National Dairy Development Board (NDDB), Annual Report 1994-95.
12. R. J. Singh, V. Prasad & S. M. Dingar, "Indian Agricultural Policy

in the Context of New Trade and Industrial Policy", Indian Journal of Agricultural Economics, Vol. 47, No.3, July-Sept. 1992.

13. World Bank, India-Livestock Review Key issues for Enhancing Growth and Development, 1995 (unpublished).
14. Indian Dairy Association, Dairying in India 1995, Export Statistics and Milk and Milk Products as made available by Agricultural and processed Food products Export Development Authority (APEDA).
15. The Hindu Business Line, "Milk Products Export Ceilings Slashed", 9th Sept. 1995.
16. National Dairy Development Board (NDDB), Annual Report 1994/95, "Operation Flood and the Dairy Sector,, pp.19.
17. NDDB, Ibid.
18. NDDB, Constraints Coming in the Way of NDDB in Achieving its Objectives (An unpublished Note), 1995.
19. "Butter Oil to be Imported for Ending Shortage", Times of India, 14th Sept. 1994.
20. NDDB, Ibid. Annual Report, 1994-95.
21. R.R. Gupta (ed.) Dairy India, 1992, pp. 112.
22. S.N. Misra and R.K. Sharma, "Demand for Milk in 2000 A.D", in Dairy India, 1992, pp. 29-30.
23. Kirit S. Parikh, "4% Growth Rate in Agriculture Over the 1990's,, IGDRI, Bombay, 1991.
24. World Bank, India-Livestock Review, May 1995, pp. 51.
25. Dr. H. Schelhaas, "Mega Trends in World Dairying", in Dairy India - 1992, pp. 42-43.
26. H. Schelhaas, Ibid.
27. R.P. Aneja, "Traditional Milk Specialities : A Survey", Dairy India - 1992.
28. World Bank, India - Livestock Review, May 1995, pp. 59.
29. P.T.I. Economics Services, "Dairying May Emerge as Major Foreign Exchange Earner", Financial Express, Bombay, 24th June, 1994.
30. Suresh Sharma, "The World Trade Organization - I & II", The Hindu Business Line, 10th and 17th June, 1995.
31. Kurien V., "We 're do-ers'", Business India, Bombay, 31st July 1995 (Interviewed by Marlc Almeida and other Press Interviews).
32. Kurien, Ibid.
33. Kurien, Ibid.
34. N.K. Chavala, "Developed Countries Create Obstacles in Milk Product Exports", Financial Express, 20th March 1995.
35. Jayshree Sengupta, "Might is Right - Guest Column, Sunday, 30 September, 1990.
36. Sengupta, Ibid.

37. Planning Commission, Government of India, Brahamprakash Committee Report on Model Co-operative Act, 1989.
38. Dilip Shah, "Co-operative Legislation in the Context of Economic Liberalization" Cooperative Perspective Jan-June, 1997.
39. Kurup M.P.G., "Embryo Transfer", Dairy India 1992, pp. 171-172.
40. Radhakant Rao, "India's Dairy Expertise much in Demand", Indian Perspective, New Delhi, March, 1995.

□□□

7

Entry of Private Dairy Operators in Mehaana Milkshed of Gujarat-Rapid Assessment

Introduction

This initial micro-level study origins from our earlier study on the policy of delicensing of dairy product manufacturing in our earlier (D. Shah, 1992) wherein we observed that expected benefits of growth, investment, technology and export on account of entry of private sector are unlikely to be achieved and thus liberalised dairy policy would be more harmful than otherwise. That study was undertaken prior to the advent of Milk and Milk Products Order (MMPO) and now since the new policy has been in operation for about four years and consequences are being gradually felt, we intende to extend the earlier study to analyse the micro-level experience during the post-MMPO period.

2. Review of the dairy policy and MMPO

On the basis of the successful Anand Pattern dairy co-operative model, The National Commission on Agriculture had recommended development of dairying through small and marginal producers and the "landless" (Ministry of Agri, 1976). By accepting the "Operation Flood Project" which in fact, was an instrument of the national policy of replication of the Anand Pattern, national planning gave it a positive recognition in subsequent plans (National Planning Commission, 1980, 1990). Reversal of this policy took place in July'91 in favour of opening up of the sector to private operators, including multinationals, with the declaration of the delicensing of the dairy processing sector which exempted the dairy processing industry from the Industrial Development and Regulation Act (1951). However, within a short period, the profit mongering behaviour of the private sector was culminated in milk crisis in the cities, excessive capacity creation and absence of level playing field for the co-operatives which ultimately led to the promulgation of MMPO (Ministry of Agri, MMPO, 1993). The private sector and also the World Bank severely criticised the MMPO as "backdoor licensing". The World Bank Study (1995) stated that, among others :

(1) The delineation of milksheds would result in creation of local mono-polies, leading to erosion in operational efficiency and lower margin to producers (2) Unrestricted competition leads to lower processing margin as firms will be forced to attain operational efficiency (3) Lower processing margin will lead to lower consumer price (4) Higher price to producers would promote production and supply which in the long run help diversification of product line of higher value added products while satisfying the fluid milk demand (5) Conditions of setting for new milksheds for new-comers is unreasonable as it will impose higher investment cost for new firms which would restrict the entry of potential entrants (6) Exclusive reservation of milksheds for the co-operatives provides unfair advantage as these milksheds are developed by World Bank resources in the form of subsidies (7) Co-operatives' argument against state subsidies to new enterprise by Harayana etc. is a "double standard" as they themselves have enjoyed such subsidies (8) The provisions of MMPO regarding dictating stock and to cease the stock of private enterprises will increase market uncertainty and will reduce the capacity for inter-regional arbitrage and (9) objective of imposing barrier to private entry on the ground of sale of contaminated, adulterated products harmful to consumers could be efficiently and effectively met by streamlining the regulatory/enforcement agencies responsible to enforce conditions of hygiene and sanitation in the dairy plants.

We shall attempt to enquire into the theoretical and empirical validity of these arguments. However, it maybe noted that in USA itself, these observations are challenged by reputed research organisations. A comparative study of public sector and co-operative dairying in India conducted by the United States International Society (Economic Times, 6 May'93) For Public Analysis dismissed the private sector argument that deregulation would result in development of the dairy sector and impart a significant investment in the farm services. Describing these arguments as "red herrings" the study said "The farmer-oriented dairy co-operatives are the only means by which farmers themselves can establish a competitive position in market". It noted that in comparison to a dairy co-operative, the private firm services to the farmers were "marginal". It has further warned "In the dairy industry, co-operatives are about to perish as a sacrifices to the "Gods of economic liberalization" then it may result in the exploitation of the agricultural community that has been the root cause of stagnation in that sector of economy. "The study also noted that dairy industry throughout the world operate under MMPO like regulations. Thus, the

World Bank argument are well replied by the USA study. The provisions of the MMPO were to regulate the process of liberalization in such a way that interest of all consumers, producers, co-operatives and even government could be protected and a kind of balanced and ordered growth could be achieved.

However, the private sector pressure tactics succeeded to tilt the policy in their favour and in May'93, major provisions of the MMPO were so diluted that the modified MMPO came to be regarded as "toothless tiger". It is viewed that the amended MMPO has opened the gates for the mushrooming tiny units in the dairy sector who are not "amenable to regulation" and "exposed" to public gauge (Tushaar Shah, 1995) and this is regarded as a major threat to co-operative sector, more difficult than to contest and compete with MMC and major dairy companies.

Though systematic data on the private dairies are not available on account of large scale entry of private dairies, big and small (in fact, such units do not follow even the mandatory obligation of providing such data to the licensing authority, we were able to informally derive some information from informal sources, as indicated) in Table -1.

Table : 1 MMPO : status as on June'95

	Coop	Others	Total
Applicants(no)	256	442	698
Existing capacity (LLPD)	235	169	404
Incremental capacity proposed (LLPD)	73	465	539
Total capacity (LLPD)	308	634	942
Rural marketable surplus (LLPD)	-	-	456

Expectedly, bulk of the applicants (63%) represented the private sector, taking best advantage of liberal entry. Note that while 58% of the existing capacity is in the co-operative sector (including government dairies), their share in incremental capacity is only 14% (the rest is in the private sector) – should all the applicants be granted license, 33% of the eventual total capacity would be in the co-operative and 67% in the private sector and, more ominously, the country's total capacity would be almost twice the total rural marketable surplus.

Given the fact that milk potential spread and institutional coverage varies across states, reference to position by states would be interesting (Table 2).

Table : 2 Region-wise number of applicants (May'94)

North	Delhi	22	East	A & N Islands	1
	Haryana	71	and	Assam	2
	HP	5	Central	Bihar	10
	J & K	3		MP	19
	Punjab	57		Orissa	10
	Rajasthan	41		Sikkim	1
	UP	174		Tripura	1
	Sub-total	373		W.Bengal	10
				Sub-total	54
South	AP	42			
	Karnataka	23	West	Daman & Diu	2
	Kerala	8		Goa	1
	Pondicherry	1		Gujarat	54
	TN	28		Maharashtra	108
	Sub-total	102		Sub-total	165
			All India 694		

The highest number of applications came from UP (which had 18% of India's milk production in 1994) and more than 50% of the applications had come from North India which had 45% share in production. By implication, the private sector prefers to operate in best production areas and avoids areas where milk production/surplus situation is challenging as in the East and South.

Review of press reports on the impact of MMPO in Gujarat (which is a stronghold of the milk co-operatives with India's largest and most powerful Federation at the state level tells the following disquieting developments :

> Some dairies in Gujarat such as Baroda and Mehsana have experienced reduction of procurement, eg. Baroda Dairy faced critical 22% procurement fall (from 2.25 LLPD to 1.75 LLPD) .... About 12-13 private dairies, operating in Baroda, have attracted distributors of Baroda Dairy by offering higher commission of 25-30 paise per 500 gm pouch against Baroda Dairy's usual rate of 7-8 paise per 500 gm of milk pouch (Economic Times, 2-1-96). Same has been the strategy adopted by the Sardar Dairy against Sumul) dairy in Surat... In Bhavnagar, co-operative procurement had halved, from 0.45 LLPD in 1992 to 0.25 LLPD in 1995 (Rajiv Shah, Times of India)... In Mehsana, where co-operatives had made decades of investments in raising productivity, the Sardar Dairy started collecting some 1 LLPD within six months (Shah 1995) of

appearance .... In the coming years, it is expected that with about Rs. 2000 crores investment, MNCs are likely to collaborate with Indian counterparts and as much as Rs. 900 crores will be invested in Gujarat (Economic Times, 1995) alone. (Thus, private sector will gain in strengthe), particularly in Saurastra where Gujarat Dairy Development Corporation (GDDC) has been declared a sick unit.

Thus, judged by the direction of the events, the shape of things to come is far from encouraging for the co-operatives.

3. Macro-level issues under investigation

In the light of the above developments in the post-MMPO period, the issues we propose to investigate could be spelt out in the fashion of conventional hypothesis of traditional research study :

1. major private dairies operate primarily in those milksheds of Gujarat which have been developed by the co-operatives and thus minimise their investment and operational costs.
2. Entry of private dairies has led to excess of capacity over marketable surplus of milk in the milksheds.
3. The private sector is dominated by small/tiny units (not MNCs or large Indian private companies) which are more difficult to monitor and compete with, on account of their capacity to avoid regulation, their scattered presence and style of operations.
4. The method of collection is not direct from producers but those from middlemen traders and contractors.
5. All evil practices (in measurement and pricing) of the traditional dairy sector have been revived to exploit producers.
6. They endeavor to weaken the co-operative structure by using various methods (offer of credit against a lien on the lactation production, advance payment etc.)
7. Much of their milk is used for product manufacturing rather than selling as liquid milk.
8. All kinds of irregularities, in collection, processing and marketing, are practised in violation of the standards.
9. Private dairies evade tax to enhance margins.
10. Urban market competition may lead to unhealthy competition, even among the co-operatives which, if not controlled, may weaken the Federation at the state level.
11. The deficiency process has already set on the co-operatives as private dairies are drawing away skilled and experienced technical staff/personnel from co-operative sector and

12. Cooperatives may not be able to withstand the crisis if the fast developing inefficiency, low productivity, lack of business culture are not set right.

4. Design of the Study and its objectives

Focus of our study will be on the milkshed of Mehsana which experienced, for the first time anywhere in Gujarat, threat to the cooperatives from a private dairy called the Sardar Dairy. Being a well-established cooperative milkshed, a number of lessons could be learnt from its post-MMPO experience which could have implications on the future of the milk cooperatives elsewhere in Gujarat as well. In the light of the central theme of our study, our approach to the issues has been rather unconventional, informal and flexible, largely conditioned by such factors as availability of data, time and resources. In the light of the issues raised earlier, we may spell out specific objectives of our study as:

1. To critically analyse the working and performance of Mehsana District Milk Producers' Cooperative Union during 1991-95 with particular emphasis on procurement and marketing dimensions and to assess the impact of the entry of private operators in the milkshed.
2. To enquire into the numerical size and structure of private dairy operators and their procurement practices.
3. To estimate investment and investment pattern of private dairy units.
4. To review critically the market strategy, conduct and behaviour of private dairies in their effort to capture market share in a major urban market (Ahmedabad).
5. To enquire into the malpractices of dairy operators to promote margin at the cost of consumers and the exchequer.
6. To undertake comparative review of the production enhancement efforts and investment of cooperatives and private dairy operators.
7. To evaluate the direct and indirect benefits drawn by the private sector out of cooperative institutional structure, investment and infrastructure and thus
8. To clearly underline the short/long term challenges of private dairy sector and to suggest appropriate short/long term policy actions to strengthen the cooperative dairy enterprises.

5. Methodology

This study was rather difficult as it attempted an inquiry into highly sensitive and business related aspects which was a rather risky

proposition. However, committed as we were to independent, worthwhile to undertake all the risks to unfold the facts).

As stated earlier, it is quite unconventional in the context of the database also. The issues are so sensitive that even the co operative dairy officials and organizations were hesitant and reluctant in their interaction with us. Limitations notwithstanding, we have done our best to lay hands on published material, unpublished reports etc which have been utilized appropriately after satisfying ourselves of their validity.

Basically, we adopted a Rural Appraisal Method with great informality at various stages with various key informants. We had prepared some broad questions to interview producers at the villages, consumers in the urban areas, senior cooperative officials etc. Such informal sessions helped us to derive facts, cross-verify the data, develop our perception and understanding of the problem. our key informative included the producers, consumers, managers, workers, cooperative leaders, technicians, academicians and even reporters of the press.

6. Limitations of the Study

This rapid study provides a basis for a larger scale research study with more systematic methodology and it cannot be claimed to be a scientific enquiry. Moreover, authentic data was the most limiting factor, our efforts to verify their authenticity notwithstanding. Our interpretations of the available facts and recommendations are purely judgemental and may appear at times academic. For another, though major private dairies are more than one, we have concentrated on Sardar Dairy only due to its significant size and peculiar growth and it was difficult to study its market behaviour in other cities of Gujarat, though we had tried to look into its behaviour in our home city of Surat. Thus, our study is highly a qualified study for the policy guidance but provides an interesting account of the micro-level scenario of post-MMPO dairy industry in Gujarat with specific reference to the Mehsana milkshed.

THE MEHSANA MILKSHED AND THE MILK PRODUCERS UNION

Profile of the Mehsana milkshed

Mehsana district (9027 sq km) has a population of 29.4 lakh, dominated by the Chaudharis, Rabaris and Patels. Relative index of development (CMIE) ranks mehsans at 73 which is less than the state's 114, reflecting Mehsana's relative backwardness. Most of the Talukas are "backward", as identified by the IG Patel Committee. It is a draught prone area with poor forest cover (1.8% area under forest, as against the State's 10% and the country's 22%). The cropped area is 43% as against 25% of Gujarat. Irrigation is much dependent on the erratic "rainfall" as no major river/irrigation system operates. "Wells" are the source of water, levels of which have gone down in the absence of regular recharging and over-use. Paradoxically, a "water scarce" area is regarded as a leading milkshed. (Part of the explanation lies in the composition of holding of land and the occupational structure of rural). People where more than 76-80% are small and marginal farmers and 65% are dependent on agriculture and allied activities. Dairying has turned out to be an assured income generating activity, transforming the face of rural Mehsana. A comparative analysis, as presented in Table : 3, shows, Mehsana as the number one milkshed of Gujarat. No wonder that it has greatly attracted private dairy operators.

Marketable Surplus

The single major factor which determines the extent to the dairy industry could develop is the quantum of milk production and marketable surplus . On the basis of milk yield and mikh animals, milk production and net marketable surplus are estimated as under:

MEHSANA MILKSHED PROFILE

TABLE 3

Dairy Economy Features	Total animals in Milk to total bovine popu.	Total Buffaloes to total bovine popu.	Cattle buffalo Ratio	% Share in Milk animals milk animals	%Share of district in total	Rank	Milk prod. share (%) 1993	Milk yield day Kg. cow	Milk yield day buffalo	1992	Rank
1. Jamnagar	35	25	75	3.80	490	13	3.83	2.65	3.13	3.13	14
2. Rajkot	37	27	73	5.45	948	07	5.96	2.71	3.42	5.76	08
3. Surendranagar	40	26	74	3.51	683	14	2.77	2.64	3.49	2.98	16
4. Bhavnagar	42	36	64	6.32	982		0.28	2.70	3.80	6.35	07
5. Amreli	37	30	70	3.85	925	12	3.87	2.65	3.86	3.93	12
6. Junagadh	39	33	67	6.89	924	04	7.50	2.56	3.95	7.03	06
7. Kutchh	43	22	78	5.10	248	08	5.47	2.73	3.63	3.42	03
8. Banaskantha	41	39	61	8.30	1056	03	9.70	3.12	4.02	8.42	03
9. Sabarkantha	40	47	53	6.99	1586	05	6.80	2.21	3.15	7.57	04
10. Mehshana	47	62	38	9.50	1287	01	13.35	2.87	4.37	13.51	01
11. Gandhinagar	51	71	29	0.99	1906		-	-	-	-	
12. Ahmedabad	45	48	52	5.17	893	06	6.30	2.50	3.76	7.13	05
13. Kheda	43	68	32	8.10	1668	02	10.16	2.37	3.57	11.50	02
14. Pachmahals 33	30	70	8.92	2168	10	4.40	1.01	2.24	5.02	09	
15. Vadodara	36	40	60	4.92	1296	09	4.59	1.93	3.07	4.80	10
16. Bharuch	37	35	65	3.04	718	16	2.42	1.70	3.03	2.36	17
17. Surat	40	36	65	4.72	1062	11	4.04	1.46	3.31	4.39	11
18. Valsad	32	26	74	3.64	1532	15	2.30	0.94	3.16	3.09	15
19 Dangs	28	16	84	0.39	689		-	-	-	-	
GUJARAT	40	39	51	100	885		100	2.35	3.58		

Source : Bulletin of Animal Husbandry and Dairying
Directorate of Animal Husbandry 1993

Ahmedabad 1992-93

Milk production, retention and marketable surplus (lakh kg/day)

Milk production (1)	Cow	1.27	
	X-bred	1.18	
	Buffalo	10.58	
	Total	13.03	
Rural consumption (2)			2.60
Village milk products			0.50
Total rural retention		3.10	
Rural marketable surplus		9.90	
Urban consumption (3)			1.20
Net marketable surplus		8.70	

(1) Source : Directorate of AH, Govt of Gujarat (1990-91)

(2) @ per cap rural consumption 115 gms/day, 22.80 lakh (1991) rural human

(3) @ per cap rural consumption 186 gms/day, 6.4 lakh (1991) urban human

Thus, a net marketable surplus of average 8.7 lakh kgs/day defines the upper limit for most efficient dairy industry in the district.

Mehsana District Cooperative Milk Producers' Union

In 1958, the UNISEF supported the establishment of a cooperative dairy under the leadership of Late Mansinghbhai Patel. The Union was registered in 1970 and started supplying milk to Ahmedabad city. Dudhsagar Dairy was installed in 1965-66. Its present installations include 1 lakh litre processing capacity along with dairy product plant, 5 chilling centres, two cattle feed plants and one Bull Mother Farm.

Unlike Kheda ("AMUL"), Mehsana Union established itself in much more favourable conditions : (1) it had a better tradition of dairying, given its agro-climatic features, as explained above. Notably, "Mehsani Buffalo" also is a major buffalo breed (2) It had the benefit of the experience of "AMUL" and received its moral support (3) It started with an assured market from Defence by way of a "buy back" arrangement for supply of milk powder— well-established marketing network of the Federation also helped it secure a market for dairy products (4) It is one of the early beneficiaries of loan and grants under Operation Flood and (5) It has done better than others in enhancing animal productivity.

Performance of the Union

Activities of a cooperative union are directed to the needs of its member-producers. Specific performance indications in this regard are discussed below (Table: 4) :

(1) Of the 1020 registered DCSs (93% of the villages), 95% are functional. A decline in the % **age of functional societies is observed** since 1993-94, though the CARG for past five years is positive (2) membership per DCS has been well above the viability norm of initial 100 members and has a positive trend. Mehsana's 33% women membership is almost double of the rest of the Gujarat unions(3) however, procurement per organised society has registered a negative CARG at -3.7% which is a matter of concern. In fact, overall milk procurement itself has been on the decline since 1992-93, more so in relation to the district's milk production, (incidentally, procurement of unions such as Kaira, Sabar, Ahmedabad, Bharuch and Ghandhinagar also declined during 1990-94) Note that the estimated production of the district has a positive CARG Of 3.1% because of increased mikh animal population though the field has declined since 1990-91 (4) While AI, animal health cover and sale of seeds have a positive trend, sale of feed has a continuous decline (5) Cost of inputs/litre have almost trebled since 1990-91 and (6) While overall sale of liquid milk has increased at a CARG of 32.7% and, as a % of procurement, has increased from 11% to 28% in last four years (indicating the Union's commitment to the liquid milk consumers), sales within the district has a CARG of 22.8%.

Table : 4 Growth indicators of the Union
(1) Institutional development

	Village society			Members				Procurement
year	Regis tered	Func tional	%	Total (000)	Per Regd society	women (000)	% women	per society
1990-91	943	886	94	270.4	286	83.0	33	293
1991-92	950	921	97	272.8	294	83.0	30	294
1992-93	994	952	96	285.4	287	93.0	33	318
1993-94	1009	963	95	289.6	297	95.0	33	297
1994-95	1020	970	95	292.8	287	96.0	33	297
CARG, %	2.0	2.3	2.0	0.1	3.7		-3.7	

(2) Milk production and procurement

year	Milch animal Nos 000	yield kg/d	Dist milk prodn (MT/Y)	procurement (000 kg) Annual	Daily	% prodn	Price : Rs/kg fat
1990-91	435	11.2	488	276200	757	56	84.3
1991-92	444	11.0	487	279765	766	57	107.5
1992-93	475	10.9	517	315729	865	61	111.2
1993-94	497	10.7	530	299501	821	57	113.5
1994-95	521	10.6	552	247222	705	46	149.6
CARG, %	4.6	-1.3	3.1	-1.8		15.4	

(3) Production enhancement

year	AI done (000) Cow	Buff	All	Animals treated (000)	BCF sold (MT)	Lucerne seeds sold (kg)	Cost of (Rs) inputs/ litre
1990-91	56	101	157	386	123	13	0.10
1991-92	113	208	322	482	123	33	0.20
1992-93	133	218	351	536	104	65	0.25
1993-94	140	139	279	537	91	56	0.21
1994-95	131	122	252	504	86	53	0.28
CARG, %	23.6	4.7	12.6	6.9	-8.5	41.2	29.4

(4) Sale of liquid milk

Year	Within district	%	Out of district	Total sales
1991-92	16398	53	14338	30736
1992-93	15658	40	23976	39634
1993-94	13642	22	47705	61347
1994-95	2.4		55.8	32.7

Sources : Dept of Animal Husbandry (Gujarat) and Annual Reports of the Union and GCMMF

Comparative analysis of cooperative vs commercial dairy

A recent study of Parthsarthy (1992) has compared the performance of the Mehsana Co-operative (MCU) with Nestle India Ltd (NIL) and Greater Bombay Milk Scheme (GBMS). The study indicates that (1) MCU consistently paid higher procurement price (2) Producers' share in the consumers' rupee varied between 68-87%

against 49-64% for GBMS but farmers received less than 12-16% of the price consumers paid for the value added dairy products of NIL (3) MCU proved farmer producer friendly, GBMS proved loyal to consumers while NIL stick to its investors interest and as a result MCU had a far more aggressive production enhancement programme than either NIL or GBMS. MCU's feed programme was far more vigorous than NIL's, the latter supplied 1 Kg of cattle feed for 11-13 litres of milk but MCU distributed 1 kg of cattle feed for less than three litres milk supplied. Similarly NIL's breeding programme provides a limited breeding resource while over 80% of buffaloes and 69% of cows owned by MCU members get services. Rise in animal productivity and substantial rise in procurement per household in the fold of MCU is noteworthy and (4) MCU also extended greater benefit to consumers – MCU charged clearly 25% less (ex-factory) on infant milk compared to NIL.

Thus, this comparative study shows clearly that Mehsana Union is capable of doing much better performance than the commercial unit. The Union is a well managed cooperative, generating a large economic return like any commercial enterprise. Since it is a very efficient organization, it is capable of having establishing "social significance" due to many social impacts. A number of studies (by VS Vyas, SM Patel, Dilip Shah, Somjee & Somjee etc) have clearly established how this Union has contributed to socio-economic development of poor producers in this draught-prone area. Our recent visits to dairy and milkshed have provided unqualified impression that in implementing numerous imaginative projects to help small, marginal and landless producers and women, the Union has surpassed other cooperatives, including "Amul". Another unique feature of the novel and distinct approach of the Union's leadership has been to integrate itself with other existing agencies and programes for rural development.

It has been implementing an unique "Progeny Testing Programme" (since 1985 and the data show that Mehsani buffalo could match with Muriah buffalo as the lactation production of female calves born to the first two proven bulls of all the three groups has been over 2000 Kg.) The farmers are encouraged to rear the female calves covered under this project. The Co-operative Development Programme(CD), orients women and small producers in scientific training and cooperative education. "Milk Day" organized by the dairy provides an unique opportunity to know all about co-operative organisation, business transactions, marketing arrangements and scientific animal husbandry practices. Animal Husbandry Assistants providing scientific extension and services (insemination, vaccination, first aid, dewarming, sample blood collection for diagnosis etc) in selected societies have been of great help to producers. Also initiated are special schemes for women, landless and poor producers in collaboration with various concerned agencies to promote effective participation of weaker sections in alleviation of their poverty. Some of these schemes are :

(1) Distribution of dairy animals to the poor through IRDP (2) Development of common fodder farm on waste-land and its distribution to landless/agricultural labourers and marginal producers (about 25 societies are active on the common fodder farms, 2807 MT of green and 38 MT dry fodder have been produced on 1210 hectares since 1994 which were sold to 2618 beneficiaries which included 929 small, 531 marginal farmers and 952 landless labourers and 204 others). Subsidised hand chop-cutters and sickles are distributed, fodder demonstrations are conducted and new fodder seeds (Harasona, SK-88001, Hybrid Jowar, African tall maize, Oak (Kent) are also distributed (3) Widow and Other women are extended cattle loan to enable them to own dairy animals, in collaboration with Social Welfare Board – Rs 1.3 crores advanced upto 1994-95 to 3700 beneficiaries of 18 societies (4). The Group Insurance Scheme ("Social Security") benefits over 1.5 lakh members (including 94343 women members) of 854 societies. Union provides 1/3 of the total premium (Rs 6.3 lakh), Insurance amount of Rs 5000 has been paid to 1364 claimants (Rs 41.7 lakh) and the upper age limit has been increased from 54 to 59 to encourage participation (5) Under the Savings Linked Insurance Scheme, so far 29922 members of 225 societies are being helped through promotion of savings and extension of Insurance benefits to the poor (6). The Group Animal Insurance Scheme covered 1.2 lakh animals in 611 societies – of the total 24769 claims, 3992 claims were admitted and Rs 1.71 crores have been paid – Union's share of the total premium of nearly Rs 20 lakh was reimbursed to the societies. Thus, efforts are being made to extend special benefits to small, marginal and landless producers.

GROWING DAIRY PRIVATIZATION IN MEHSANA MILKSHED

1 Review of MMPO applications in Gujarat

It is attempted to draw a brief profile of some selected private dairies operating in Mehsana milkshed. The private dairy operators in Gujarat have been trying to enter the area of dairy product manufacturing since 1991, though the initial process was rather slow. Table : 5 shows the current status in this regard in terms of MMPO applications. It could be observed that :

(A) Out of total 107 applicants, as many as 81% are from private sector and they could be identified as (1) old dairy operators in the major markets like Ahmedabad (2) Traditional contractors or Bhaiyas involved in the business – commission agents, Gamadiwala, Gayatri, etc (3) old anti-cooperative lobby like the Polson (4) None belongs to any known corporate enterprise or MNC, most of them have no specific identity in the market and are so tiny that it is difficult even to locate them, far from monitoring them (5) Former cooperative employees (6) young dairy engineers and entrepreneur jumping into the fray to reap benefits of new economic policy and (7) Most of the owners of private dairies happen to be Patels.

Table : 5. MMPO Applications in the district of Gujarat

	District	No of Applicants		Rural Marketable surplus	Maximum Capacity (LLPD)			Deficit
		Coop	Pvt	(LLPD)	Coop	Pvt	Total	(LLPD)
1	Ahmedabad	1	8	2.13	2.00	8.82	10.85	-8.72
2.	Amreli	1	1	0.86	3.00	0.40	3.40	-2.54
3.	Banaskantha	1	9	3.54	4.00	7.90	11.90	-8.36
4.	Bharuch	1	4	0.85	0.60	3.90	4.50	-3.65
5.	Bhavnagar	1	1	1.93	3.60	0.00	3.60	-1.67
6	Gandhinagar	1	7	0.55	10.30	6.45	16.75	-16.20
7.	Jamnagar	1	10	0.92	0.20	0.00	0.20	0.72
8.	Junagadh	2	1	2.69	3.00	0.50	4.50	-1.81
9.	Kuchh	2		1.14	3.60	0.00	3.60	-2.46
10.	Kheda	1	12	6.53	14.50	12.73	27.23	-20.70
11.	Mehsana	1	16	9.62	10.05	17.18	24.10	-14.61
12.	Panchmahal	1	4	1.32	2.00	4.70	6.70	-5.38
13.	Rajkot	1	0	1.7	1.00	0.00	1.00	0.70
14.	Sabarkantha	1	14	4.56	9.50	12.35	21.85	-17.29
15.	Surat	1	1	2.8	4.00	1.00	5.00	-2.20
16.	Surendranagar	1	3	1.39	4.00	2.90	6.90	-5.51
17.	Vadodara	1	5	1.66	3.00	4.50	7.50	-5.84
18.	Valsad	1	0	0.92	1.00	0.00	1.00	0.80
19.	Dangs	0	0	-0.19	0.00	0.00	0.00	0.00
	Total	20	96	44.92	79.35	83.33	160.58	

Note : Coop includes Government managed dairies

(B) Spread of private dairies is not balanced. Districts like Dangs, Valsad, Bhavnagar, Kutch, Rajkot and Jamnagar receiving low priority. Evidently, the concentration is in those milksheds which had been developed under Operation Flood by huge investment and institutional development – they have mainly identified their operational areas in the developed cooperative milksheds of Mehsana (16), Sabar (14), Kheda (12), Banas (9) and Ahmedabad (8). Also, their proposed milkshed often spreads over more than one district —es. Sardar Dairy, registered in Mehasana, has shown its milk sources in Mehasana, Ahmedabad, Banas and Sabar.

(C) Unlike the cooperatives, who opt for large capacity, private units do not exceed 2 LLPD, many of them have asked for even smaller capacity of 0.20 LLPD

(D) Finally, what could be the feasibility of the subsistance of these capacities in the context of the milk supply and the state of marketable surplus in the area ? Based on certain assumptions on yield/animal, rural/urban consumption etc, one estimate suggests that :

(a) In most of the districts, there will be "deficit" with installed capacity exceeding marketable surplus (except Rajkot and Jamnagar) and

(b) the "deficit" would be alarming in the developed areas of Kheda, Mehsana, Ahmedabad and Sabar.

However, except for two tiny private dairy operators, none of the private dairy units have so far been granted any license.

Table : 6. Registered location of private dairies in Mehsana

	Name of the Applicant	Applied capacity (LLPD)	Location
1.	Mehsana Dairy	10.05	Mehsana
2.	Vimal Dairy Products	2.00	Mehsana
3.	Vishwas Milk Centre	0.20	Vijapur, Mehsa
4.	Avkar Dairy	0.02	Vijapur, Mehsa
5.	Casco Food	1.50	Mehsana
6.	Sundhiwala Dairy	0.35	Kheralu, Mehsana
7.	Sardar Dairy	1.00	Mehsana
8.	Anmol Dairy	2.00	Mehsana
9.	Sughad Dairy	0.35	Gandhinagar
10.	Rich Dairy	0.50	Vadgam, Banas
11.	Polson Dairy	2.00	Chandkhed, Ahmedabad
12.	Gomadiwala Dairy	2.00	Ahmedabad
13.	Satyam Dairy	1.00	Gandhinagar
14.	Zarf Dairy	0.35	Ahmedabad
15.	Shrinathji Dairy Farm	0.60	Umreth, Kheda

16.	Samrudh Dairy Products	0.50	Gadhinagar
17.	Gayatri Products	0.50	Ahmedabad
	Total	24.92	

Mehsana has attracted largest MMPO applications [(about 15% each of Gujarat's 107 applications and 160 LLPD of proposed capacity)]. Table : 6 shows though only half of the applicants belong to Mehsana, most of them would depend on other districts for the purpose of milk collection. If all the units are granted license, the overall capacity (24.10 LLPD) would far over-shoot the district's marketable surplus (9.62 LLPD).

3. Locational spread of private dairy operators in Mehsana

In late 1992, Sardar Dairy (2 km away from the highway, near Palvashana village) came into existence, followed by Anmol Dairy (1995) and many other operators (suppliers/vendors), aspiring to become independent "dairy plants" with product making capacity.

Table 7. Taluka-wise spread of dairy units in Mehsana

Taluka	Name of the private dairy operators
patan	Radha
Sidhapur	Rich India
Visnagar	Sughad, Gayatri, Bhagya laxmi Mahakali, Kapur
Mehshana	Sardar, Vimal, Sanjay, Gujarat
Chanasma	-
Kheralu	Mahakali, Sundhiwala
Sami	-
Haris	-
Kadi	Anmol, Sarita, Gayatri
Kalol	Vipul, Ganga
Vijapur	Gayatri, Avakar, laxmi, Suraj, Bhuvaneswari, Sidheswari, Umia Kishan, Suvarn

The above list of private dairy operators was prepared during our field visits in consultation with key informants and it covers most of the significant dairy operating units.

Table 7 gives the taluka-wise procurement of MCDU in Dec'96 which we may use as "proxy" to production spread in the talukas and which shows locational choice of private operators. Nearly 75% of the units are located in best milk pockets of MDCU (accounting for 60% of its milk procurement). In fact, almost 50% units are located in Vijapur-Visnagar area where MDCU has made a remarkable investment in these areas while the relatively deficient areas are untouched by the privates — eg, Chanasma, harij, Sarva and Patan. This demonstrates that private dairy units, without loss of time and resources, want to harvest the benefits of co-operative initiative.

On the basis of field visits, observation and selected interviews we present below a broad profile of the major private dairies.

4.1 Sardar Dairy

It came into existence as a private limited company in Oct'92. It was started by one Shri MP Patel, a dairy technologist and an ex-employee of a renowned co-operative dairy of the area during 1968-77. He is supported by other two ex-employees of the said co-operative dairy. The entire equity capital is owned by MP Patel. Prior to floating equity issues for the Public in Oct'95, it was engaged in milk processing and manufacturing of value added dairy products like ghee, butter, cheese, Shreekhand, flavoured milk etc. The present processing capacity is 0.8 LLPD. Its main interest is to produce pasteurised milk, butter, ghee, cheese, flavoured milk etc for which liquid milk is procured from various villages of Mehsana, Banas and Sabar districts. With 1.6 LLPD capacity after Public Equity issue, it would be able to develop procurement network in other districts like Ahmedabad, Gandhinagar, Surendranagar and Kheda. According to one source, it has, in fact, an installed capacity of 1.0 LLPD and produces different kinds of products

Table 8.Output structure Table : 3-6 Financial indicators

(Rs. in lakh)

Item	Quantity (Tons)	Item	31.3.93	31.3.94	31.3.95
pasteurised whole and standard milk in pouch	5675	Sales	973.77	1579.13	2544.97
		Other income	1.03	2.76	5.13
pasteurised toned milk	3660	Profit before tax	0.93	59.95	45.04
Case in	31				
Srikhand	365	Profit after tax	0.93	48.57	42.42
Khoa	365	Share capital	12.50	12.50	354.00
Butter	2000				
Milk powder	2190	Reserve and Surplus	0.93	46.37	4.99
Ghee	1000	Book Book value	10.74	47.10	10.14

(Table : 8). its procurement system is stated to have the following features according to one source : average lean procurement is 0.7 LLPD and, in the flush, 1.0 LLPD is easily possible to handle with the available installed capacity as its milkshed spreads over 12640 sq km which spread across 1936 villages (Mehshana-1119, Ahmedabad-183, Banas-287, Sabar-268 and Gandhinagar-79). However, it is procuring milk from 227 villages of Mehsana district where "Milk Producers Associations" have been established. Payment is made through bank in the name of two members (agents). Milk is paid at the rate of Rs 127/kg fat. The Associations charge Rs 7-10/Kg fat, the rest is distributed among producers. It also arranges for the supply of cattle feed the quality of which could not be ascertained. It is claimed that it has already established a network of 12000 milk producers. Dairy runs its own vehicle on fixed routes to collect milk. (It sells liquid milk in Surat, Himmatnagar, Ahmedabad, Kadi, Kalol, Visnagar markets milk in pouch and in bulk.) Products are distributed in all major cities of Gujarat through dealers on commission. Financial performance (Table : 8) in view of the short span, it is very difficult to evaluate but one would note the declining profit in 1994-95. Future plan of Sardar Dairy, with extended authorised share capital of Rs 5 crores is to expand its present capacity from 0.8 to 1.6 LLPD at Rs 425.1 lakh to be financed through promoters contribution of Rs 237.5 lakh, Public issue of Rs 115 lakhs internal accruals of Rs 38.1 lakh and unsanctioned loan of Rs 33.5 lakh. The raising of investment through issue was also to meet the margin money needs for working capital. Based on highly optimistic assumptions in growth of milk production, capital utilization and prices of milk/products, it has drawn a rosy picture of projected sales, profits, earning per share etc for 1995-96 and 1997-98.

4.2 Anmol Dairy

This Dairy at Kadi, on Kalol Road, was started by one CK Patel in June'92, with an investment of Rs 85 lakh through bank loan and personal investment (35%). The plant with 80000 litres of installed capacity, has utilization averaging 20000-30000 lpd.

Milk is procured from 25 villages, highest quantity is procured from "Joda Adraj" and lowest from "Jivapur". The collected through contractors and agents. To attract the producers, it always pays paise 12-13/litre more than the co-operative, offers cattle loan (at subsidized rate of interest) upto Rs 8000 and some subsidized veterinary facility. Most of sales is in Kadi market in products form. Of the 30000-35000 milk collected per day, 2000 litres milk is directly sold unpasteurised, about 1000 litre is sold as ghee, butter, Peda, Shreekhand and buttermilk. As for economics, the owner stated that milk is collected at an average of Rs 16/litre (10% fat) and, after processing and manufacture, an average of Rs 22/litre is earned (25% profit).

4.3 Avkar Dairy

Located in Vijapur (Vijapur - taluka), it is run on partnership of four of a Patel family (one being a dairy technologist). It was earlier only a chilling centre, presently handles only 10000 litres. The investment of four partners is of Rs 35 lakhs. The unit has no powder manufacturing capacity. However, it claims to be having a capacity of 15000 litres and proposes to expand to 25000 litres, when the SMP plant becomes operational. It has applied for registration under MMPO. It is reported that hygiene is non-existent in the unit, debris are all around as an old house was converted into a processing and packaging hall. The cleaning of equipment, sanitation and hygiene situation at the plant is learnt to be extremely poor. It is observed that it can at best handle 25000 litres of milk after expansion but its dream of making powder in existing set-up is rather distant. It buys milk from small traders/agents, without direct involvement in collection. Its milkshed areas include villages of Sabar (1359), Mehsana 1089) and Gandhinagar (75) though it has arrangement to transport milk from 20 villages only. It provides no technical inputs etc but pays Rs 18/kg fat more than the MCDU. It sells mostly milk in Ahmedabad and Bombay through distributors with commission at paise 30/litre.

4.4 Gayatri Dairy

It was started in 1994 by a milk vendor who was collecting milk on bicycle from the villages along the Vijapur-Mehsana highway. It has now a plant worth Rs 85 lakhs in which the owner (Kanakbhai Patel) has invested his own Rs 24 lakh. It has a 15 MT capacity plant though current utilization is not more than 7000 litres a day. Milk is collected from a net-work of 88 private societies in 11 villages, maximum quantity is procured from Pratapura and least from Amrapur. It pays Rs 16/litre for 10% fat (Rs 160 per kg fat). Milk collecting is done through 11 tempos which are made available to the agents. Milk sales is done directly to Ahmedabad through agent, at a commission of Rs 3/crate. The annual cost of sourage is Rs 75000 and processing cost is Rs 1.5 lakh and extension cost is Rs 75000. To attract milk producers, paise 25-30/litre is paid more than the co-operative price. Emergency credit, social credit, animal purchase credit, bonus and gifts are given once in a year to the producers.

4.5 Sundhiawala Dairy

It is the only significant private unit in Kheralu (commissioned in July'91), one of the finest milk pockets. it is managed by partnership company of PM Desai and HB Desai. It has also applied for MMPO license. The installed capacity of the plant is claimed to be 27,000 litres of raw milk processing, of which 10,000 litres of processed milk is sold as a whole, toned or standard milk. The unit also gets SMP manufactured at an outside plant. It is claimed that it could manufacture 34 MT of SMP in this manner. (It also proposes to produce 20 tonnes of casein. However, it is believed that presently it handles 15000 litres (though Sardar villages of Mehsana (48), Banas (7-8) and Sabar (417), around 25000 litres and has some 1.4% share in total collection) though it could handle upto 35000 litres.) It cannot produce ghee and SMP even occasionally as it has very poor sanitary and hygienic conditions and its surrounding is full of debris. It markets about 12000 LPD of milk to Ahmedabad and sends a tanker to Bombay (800 litres), and Mount Abu (600 litres).

5. Concluding observations

One could, at the end of this discussion, reflect on the following aspects :

- Major private dairy is Sardar Dairy followed by tiny units.
- Their concentration is in the operational area of the Union.
- This high concentration of private dairies is explained by the fact that privates as expected are never interested in the backward areas like Sami/Harij, but their strategy is to take best advantage of the best developed milksheds of Mehsana Dairy which has made a lot of investment in production enhancement.
- Finally, from the point of view minimum hygienic standards, most of these units can not make any positive claim.

MILK PROCUREMENT : RELATIVE SHARES OF PRIVATE DAIRIES – THEIR CONDUCT AND PRACTICES – RESPONSE OF THE MILK UNION

This section provides alternative estimates on share of the procurement by competing agencies in Mehshana, discusses the conduct and practices of the private dairies and the response of the producers and that of the co-operative milk Union to such development. Since estimated shares of procurement were obtained from three different sources, care has been taken to judge their validity.

Alternative procurement estimates

(a) Union's estimate

Based on its internal sources (societies and supervisors), the Union has estimated the total procurement by some 11 private dairy operators (Table : 9) at 1.7 LLPD or about 19% of its own procurement. This seems to be an underestimate, compared to the alternative estimates (especially of the Sardar Dairy). Further, it clubs a large quantity of procurement (above 60) under "others" category. Little is known about "Kishan" which collects a significant 7% of total private collection next to Sardar's 12%. Note that, with few exceptions, most of the operators are confined to a few talukas for collection of Milk.

(b) Sardar Dairy's estimate

Sardar Dairy estimates (Table : 10) total milk production of Mehsana at a very high of 25 LLPD (as against Federation's 12 LLPD and alternative estimates of 13-15 LLPD, including the Departmental estimate) and it mentions 35 dairy units (many of which do not belong to Mehsana) but excludes the unorganised section of contractors, traders etc. It puts its own procurement at 0.95 LLPD which is five times to the Union's estimate.

(c) Federation's estimate

The Federation's estimate (Table :11), based on per cap rural retention at 200 gms, puts private procurement at around 20% while the co-operative share is taken at 60% approximately.

Table 9 MCDU Estimate of private dairy procurement (kg/day) in December, 1995

Taluka	Sardar	Vimal	Anmol	Rich	Gyatri	Sanjay	Kheralu Mahakali	Best Sundhia	Kapoor	Kishan	Others
1 Patan	5655	2550	-	730							5240
2. Sidhapur	1536	1682	-	1560	-	-	200	-	-	-	3953
3. Visnagar	1605	150	-	-	3600	2175	-	-	-	-	14690
4. Mehsana	5180	755	-	-	1340	-	-	-	-	-	14040
5. Chanshma	2690	2500	-	-	-	150	-	-	-	-	4415
6. Kheralu	55	130	-	3785	-	-	2285	1772	933	-	-6683
7. Somi											
8. Harij	1310	500	-	-	-	-	-	-	-	-	1000
9. Kadi	2142	75	1378		210	-	-	-	-	-	15400
10. Kalol	-	-	-	-	60	-	-	-	-	-	14858
11. Vijapur	-	-	-	-	-	-	-	-	-	12450	28352
Total	20173	8342	1378	6075	5240	2325	2485	1772	933	12450	108631
	(11.87)	(4.31)	(0.81)	(3.57)	(3.08)	(1.36)	(1.47)	(1.04)	(0.54)	(7.33)	(63.96)

Source : Collected data from Mehsana Dairy and compiled and processed by us.

TABLE 10
SARDAR DAIRY'S ESTIMATE OF PRIVATE DAIRY PROCUREMENT IN MEHSHANA

S.No.	Name of dairy	Milk procured in litres	
1.	Dudh Sagar Dairy	10,85,000	(62.28)
2.	Sardar Dairy	95,000	(05.48)
3.	Vimal Dairy	40,000	(02.29)
4.	Gayatri Dairy (Ahmedabad)	35,000	(02.00)
5.	Shresth Dairy	25,000	(01.43)
6.	Avkar Dairy, Vijapur	30,000	(01.20)
7.	Maheshwari Dairy, Vijapur	10,000	(00.57)
8.	Suvarn dairy, Vijapur	15,000	(00.86)
9.	Laxmi Dairy, Vijapur	20,000	(01.14)
10.	Umiya Dairy, Vijapur	10,000	(00.37)
11.	Sanjay Dairy, Dabhla	30,000	(01.72)
12.	Gujarat Dairy, Mehsana	24,000	(01.37)
13.	Shyam Dairy, Kukarwada	20,000	(01.14)
14.	Samruddh Dairy	15,000	(01.86)
15.	Nice Dairy	7,000	(00.40)
16.	Mahadev Dairy	8,000	(00.45)
17.	Gaytri Dairy, Visnagar	9,000	(00.51)
18.	Sundhiawala Dairy	25,000	(01.43)
19.	Rich India Dairy, Siddhpur	35,000	(02.00)
20.	Mahakali Dairy, Kheralu	10,000	(00.57)
21.	Kapoor Dairy, Vadnagar	20,000	(01.24)
22.	Adarsh Dairy, Visnagar	10,000	(00.57)
23.	Umiya Dairy, Visnagar	10,000	(00.57)
24.	Royal Dairy, Visnagar	20,000	(01.14)
25.	Taljabhai Rabari	8,000	(00.45)
26.	Radhe Dairy, Patan	10,000	(00.57)
27.	Anmol Dairy, Kadi	10,000	(00.57)
28.	Sughad Dairy, Gandhinagar	10,000	(00.57)
29.	Govindbhai Mudarda	10,000	(00.57)
30.	Maneklal, Ambaliyasan	10,000	(00.57)
31.	Mathurbha Mantri	8,000	(00.45)
32.	Kalal	10,000	(00.57)
33.	Gagar Dairy, Adalaj	8,000	(00.45)
34.	Other small Dairies	20,000	(01.24)
35.	Ahmedabad (Kalupur) Market	30,000	(01.72)
	Total	17,42,000	
	Domestic Utilization	7,46,571	
	TotalMilk Production/Day	24,88,571	

Table 11

Federation Estimate of Procurement/Production share of private dairy share in Mehsana milkshed area

Ref year	Milk producers (000 nos)	Avg Prodn. MT/day	Gross retention Mt/day	Avg surplus Mt/day	Avg Proc. by coop Mt/day	Avg Proc. by private	% of coop	% of pvt share	Retention
1991-92	273	1336	273	1062	765	297	57	22	20
1992-93	285	1416	286	1130	867	263	61	19	20
1993-94	290	1452	290	1162	822	343	61	23	20
1994-95	293	1175	293	882	706	171	60	16	26
1995-96	293	1293	293	1000	176	224	60	17	23

Source : CDD/GCMF, 1996

(d) Study assessment

In our assessment, based on in-depth discussions with key informants and field observations, private dairy units in Mehsana are of three major types (i) Major private dairy "Sardar" (ii) Some 30-35 smaller units and (iii) Private traders/contractors/vendors/separators/ Mavawalas etc.

Sardar's proclaimed procurement of 0.95 LLPD seems to be correct (Union's estimate in this regard is a gross underestimate). Allowing another 10-15% share to the smaller units, the combined procurement of organised private would exceed 1.0 LLPD while the unorganised private traders would account for another 1.25 LLPD. Thus, total private collection seems to be around 2.25 LLPD which may be around 24% of total procurement of the Union in Dec'95. This estimate is comparable with the Federation estimate. Our estimate (Table : 12) reconciles the very high estimate of Sardar Dairy and very low estimate of the Union and is close to the Federation's estimates.

Table : **12 Study estimate of private procurement in Mehsana (l/d)**

Taluka	Sardar	Sardar + other	Unorganised	Total
Vijapur	34285	-	-	34285
Visnagar	650	13692	9054	49581
Mehsana	4256	9505	7325	21086
Kheralu	14620	-	-	14620
Kalol	17249	18165	8255	46300
Siddhpur	1817	6410	4575	12802
Chansham	3952	8834	3083	15869

Kadi	13100	-	-	13100
Patan	5655	3250	5240	14145
Haris	-	-	-	-
Sami	-	-	-	-
Total	96184	63106	37532	223007

Trade practices - networking — the patrons

This section is based on our discussion with Union employees, ex-employees, milk trader, private milk operators, milk producers and informed persons at the village level. Some information were also collected through the schedules which were prepared to collect the data on private dairies.

It has always been an impression that private dairy operators depend upon the traders and agents to get their milk. The traditional system exists even now, but remarkable changes have taken place in the villages of Mehsana. We saw systematic and regular collection centers run almost like co-operatives. Even, there were villages where three centres of three private agencies were working, one can observe the intensity of competition to collect village milk as the producers now have as many as four options to sell their milk. Sardar, which has the largest private procurement network covering major parts of milkshed, claims that it has developed "customers associations" akin to co-operative societies. It may noted that Sardar Dairy considers its producers also as "customers".

Our field level inquiry suggests that there are three types of producers emerging (1) those who are exclusively loyal to Sardar (2) those who are supplying milk to Sardar as well as other private dairies and (3) those who prefer vendors, separators, Bhattawalas and other traditional operators. Our estimate is that there are about 675 societies and centers and 32000 producers (half of whom are also attached to co-operative societies) working as a part of private procurement network. For example, we were told that in one chilling centre area, Sardar alone has network of 44 villages where some 1300 members supply 5600 kgs everyday. In the same area, Vimal has 21, Rich has four and Gayatri has one centre.

The Union's Board of Directors and Chairman are unanimously elected. Motibhai Chowdhari (Chairman) has an imposing public image due to his character and honesty, austerity and simplicity of life as a true follower of Gandhiji. Thus, one would have thought that in such a co-operative stronghold, private dairy network may not pose any immediate threat. However, the following sections are working for private milk collection in villages — educated unemployed, ex-employees of the dairy, suspended co-operative members, members who left the society (for genuine grievances), non-village residents, political workers and non-chowdhari or Patel.

"Why do you support the private trade ?" We asked this question to people who included real member-producers and active co-operative supporters as well as strong sympathisers of private dairies. Their reactions could be summarised as follows :

(1) Main reason of the growing private sector dairying is not its own efficiency, but inefficiency of the co-operatives. Some producers did not remain loyal to co-operative simply because there were irregularities, dishonesty, disregard, discrimination, indifference to complaints which diverted them to private operators as soon as the opportunity came.

(2) Some producers have taken opportunistic stand, they want to reap benefits of both private and co-operative marketing.

(3) Some are attracted as the agents of private centres provide a variety of credits and advances against future milk sales. This tieing up of milk with credit is really exploitive as the producers have no choice. For example, Anmol offers animal loan upto Rs 8000, Vipul provides as much as Rs 25000 to society/centre as advance for distribution of among needy producers against committed milk supply. Persons who run Kalyana centre (Jaganpuri) conceded that producers are provided with an advance of Rs 10000. Thus "credit" is used as an instrument to attract the producers.

(4) Agents of the collection centres are provided with some commission – some at 7% of the value of collection, some at around Rs 15/kg fat, while some offer seasonal rates (summer - Rs 15, winter - RS 5 and monsoon - Rs 8/Kg fat).

(5) Agents are also provided with capital, some light furniture and equipment and some advance money.

(6) Most attractive instrument is the procurement price which in all seasons is atleast Rs 2 more per kg fat than the co-operative's. The price varies between Rs 150-170 for 100 kg fat. Some like the Srijee claim to have paid even Rs 20 more per 100 kg fat (Rs 170).

(7) Some producers prefer private operators as they collect milk at the convenient time, do not follow tight schedules of collection and accept milk at late time and also are able to obtain clarifications on standard and quantity of milk in case of complaints.

(8) Private dairy operators imitate co-operatives by providing a

symbolic assistance in occasional supply of cheap cattle feed and pay a leap service in production enhancement (though at an in-built cost to the producers).

(9) We were told that the agents adulterate milk with water (though large scale, Haryana practice of synthetic milk has not yet taken place.)

(10) A section of the opinion leaders in Siddhpur advocated private dairies on the ground that they will compete with the co-operative and provide an effective alternative in the event of any discontent. They illustrated from their past experience that once some of the member-producers, who had gone on strike demanding a price rise, were victimised and their milk was not accepted. "Gone are the days of the Union. It will have to improve its efficiency to provide better price. There has to be some opposition".

(11) Finally, socio-political factors have also their contribution in promoting private trade, originating in Chaudhari-Patel and Congress-BJP rivalry.

Response of the Union

It would be interesting to review how the Union responded to the emergence of organised private sector in the milkshed.

1991-92 Annual Report of the Union recognised "Delicensing" of the dairy industry as matter of great concern. "Under delicensing policy, any company can buy milk from anywhere, convert milk and sell the products in the market. The co-operative unions such as ours, have spent crores of rupees to implement projects to increase the productivity of our milk animals. But now, the companies, who have not spent anything, will reap the benefits of buying milk from developed milksheds". (We have seen, how this has come true — concentration of private operators is in the most developed areas of Vijapur, Kheralu and not in the less developed Sama and Haris). The Report further observed that, through the milk co-operative organization, the farmers have freed themselves from the shackles of the middlemen and warned its producers that the main aim of the private sector is profits. "To achieve this aim, middlemen will again start operating and cause division among us at the society level. If we will not be alert and careful against such moves, they will tempt us and get the benefits of our investment of crores of rupees. We shall be able to swim across the great waves of privatization in the dairy industry, if we remain united in the true co-operative spirit". It also pleaded to the government to formulate new policy (MMPO !) which would protect the farmers against exploitation.

The 1992-93 Annual Report remained silent but the 1993-94 Annual Report summarised the real consequence of the private operations. It stated that "some members of our village societies who were supplying their milk to the private traders continued to avail of the facilities of cattle treatment, cattle feed, insurance etc. thereby increasing the financial burden of the Union, inspite of repeated reminders. It is a sad fact that we have not been able to curb such activities. We should take necessary steps to stop the use of such facilities by the members who give their milk to the private traders and revoke their membership". Thus, it was accepted that there were members who were disloyal. However, the above reaction was on financial ground and not for any ideological consideration.

The 1994-95 Annual Report, for the first time, acknowledged a 14% reduction in procurement due to private dairies. It observed that "we must accept the Those fact his few of our members were tempted to give their milk to private traders, causing further reduction in our milk receipts." The Report look into the reasons why do members behave like this. "Even after joining the co-operative movement since many years, we have not been able to understand its working and our role. As a result, we have not been able to evaluate correctly the practice of private traders. A private trader looks after own interest in business, while co-operative organization is for all its members. We are all in this business together. If our faith and understanding of our co-operative organization is "matured", we can easily overcome any "crisis" and tests that we may face".

Quite a few positive steps have been taken by the Union to cope up with the situation. The procurement division have been strengthened. Better price is followed despite difficult operations. The technical inputs distribution system is being improved for easy access to producers.To strengthen the understanding of the members, systematic and intensive efforts are being made to promote cooperative education under "Cooperative Development Program" under the guidelines of the NDDB and the Federation. Special training-cum-seminar was organized at Anand for the Board of Directors. An intensive effort is being made to highlight the role and benefits of co-operatives through "Video Rath" a mobile video system as a part of the extension activity. Of late, the Federation has started the News letter "Patrika" which is circulated to all the village societies to enhance understanding of co-operative principles and to create awareness among the members on the activities of private operators. Finally, efforts are being made to formalise the membership of non-member suppliers.

Alongside, certain deterrent actions have also been taken to enforce discipline among members (though the leadership is very cautious and careful in this regard, for any likely adverse repercussions). Producers supplying milk to private dairy are given "Notice" two or three times as an initial warning. If it does not bear result, the society reports to the Union so that services and input supply are withdrawn. As a final step, society undertakes procedure to revoke membership of the producers (though the ultimate approval has to come from District Registrar of Cooperatives which is not always easy). However, these disgruntled members are the ones who become instrumental in promoting private operations and, being local faces, they work effectively also.

MARKET SHARE OF PRIVATE DAIRIES AND THEIR BEHAVIOUR & CONDUCT WITH SPECIAL REFERENCE TO SARDAR DAIRY

This Section, considering Sardar Dairy as an illustrative case, discusses the market behaviour of private dairies in major markets of Ahmedabad and Surat.

Market share of private dairies

Four major changes have taken place in milk marketing in Gujarat, the marketing challenges and opportunities are to be understood in the context of these major changes in the post-liberalization period.

(1) (Reportedly, co-operatives' liquid milk market share (Amul, Uttam and Abad, taken together) in pouch milk in Ahmedabad is decreasing at the cost of the private – share of co-operatives, 70% in 1992-93 has decreased to 62% in 1993-94 while that of private sector has advanced from 30% to 38% (IRMA, 1996).

(2) (Interestingly, while milk sales of Federation, as % of its procurement, has increased from 40% to 51%, its (1994-95 Annual Report notes that it has successfully increased the sell of Amul milk by 18%), as many as 22 private dairies in different brand names have cropped up in Ahmedabad, selling an estimated quantity of 1.94 LLPD (Table : 13.) The major ones with more than 10% share in total private milk sales are merely three – Gayatri (12%), Royal (13%) and Sardar (13%). What is significant is the remarkable change in market structure, a slow but significant shift from a sellers market to a buyers market. Estimated share of private dairies in other cities is shown in Table : 14 and Table : 15 shows the distribution of Sardar's sales by towns :

In Surat, there are two major Sardar dealers in two parts of the city (another one is dealing only in Sardar products) who together sell about 20000 litres of milk daily. Sub-dealers are many, who are paid

Rs 5/crate of 20 litres. However, Sardar had an initial entry with 30000 litres daily in view of supply gap in the areas of Bhestan and Sachin where Sumul (Anand pattern Milk Union of the district) could not develop its network and which were covered by local traditional cooperative "Choryasi" dairy. Hence, immediate loss on account of the entry of Sardar was not to "Sumul" but to Choryasi. However, short term trade techniques, unstable supply and quality did not help Sardar to

Table 13 MARKET SHARE OF PRIVATE DAIRIES IN MILK MARKET OF AHMEDABAD

Sr. NO.	BRAND NAME	PACKER'S NAME	% OF MARKET SFARE	DAILY MILK LTR. (APPROX.)	QUALITY TYPE
1.	Gayatri	Gayatri D.P. P.Ltd. Khokhra, Ahmdabad	12.38	25000	Pouch pack
2.	Royal milk	Madhav Dairy prodn. milkand, Gandhinagar	12.88	25000	Pouch pack
3.	Super milk	Satyam Dairy Sector-28, Gandhinagar	2.57	5000	Pouch pack
4.	Shreshth milk	Gamdiwala Dairy, Kankaria, Ahmedabad	9.79	18000	Pouch pack
5.	Gokul milk	Gokul Dairy, opp. V.S. Hosp. Ahmedabad	1.03	2000	Pouch pack
6.	Sampurna	Ambavadi Dairy, Ahmedabad	1.03	2000	Pouch pack
7.	Sughad milk	Sughad Dairy P. Ltd. Dist. Gandhinagar - 382 424	2.06	4000	Pouch pack
8.	Suvama milk	New Suraj, Vijapur - 382 870	3.60	7000	Pouch pack
9.	Best	Sundhiwala Dairy, Kheralu - 394 325	2.57	5000	Pouch pack
10.	Samrudh	Samrudh Dairy Pvt. ltd. Sarichej, Ahmedabad	5.15	10000	Pouch pack
11.	Sardar	Sardar Dairy Pvt. Ltd. Mehsana	15.88	25000	Pouch pack
12.	Dairy field	Dairy Field Ltd. Chandala Dist. Ahmedabad	7.73	15000	Pouch pack
13.	Amar	Anmol Dairy Pvt. Ltd. At kadi Dist. Mehsana	2.57	5000	Pouch pack
14.	Vimal	Vimal Dairy Pvt. Ltd. At. Mehsana	2.57	5000	Pouch pack
15.	Ankur	Jay Khodiar Dairy Pvt. Ltd. Adhmedabad	2.06	4000	Pouch pack
16.	Amrut	Avkar Dairy Pvt. ltd. At. Vijapur	6.18	12000	Pouch pack
17.	Nice	Genuine Commodities Development, At. Mehsana	1.54	3000	Pouch pack
18.	Sulabh/Gopala	Jogeshwari dairy Pvt. Ltd. Rakhial Ahmedabad	2.57	5000	Pouch pack
19.	Sangam	Sangam Dairy Pvt. Ltd. Ahmedabad	1.03	2000	Pouch pack
20.	Motera	Motera Dairy Pvt. ltd. Ahmedabad	1.03	2000	Pouch pack
21.	Anmol/Kamavati	Zarf Dairy pvt. Ltd. Ahmedabad	5.15	10000	Pouch pack
22.	Radhe	Radhe Dairy Pvt. Ltd. Ahmedabad	1.54	3000	Pouch pack

Source : GCMMF, 1996.

Table : 14 Market share of private dairies in major cities

	Estimated sales/day (litres)	%
Ahmedabad	104 000	38
Baroda	100 000	36
Surat	20 000	7
Rajkot	50 000	18
Sabar	2 000	1
Total	276 000	100

Table : 15 Distribution of Sardar's milk in cities

	Estimates sales/day (litres)	%
Baroda	20 000	40
Surat	4 000	8
Ahmedabad	25 000	50
Rajkot	750	1
Himmatnagar	500	1
Total	50 250	100

Source : GCMMF, 1996

retain the market. In fact. according to market survey, most of the consumers who buy Sardar milk are those who could not collect milk from Sumul agent and, of late, some casual buyers at the private shops. It is learnt that in Surat, there is a specific class of customers ("Bharwads") who are vendors themselves and who regularly use Sardar milk to adulterate of their milk. Their choice of Sardar milk is due to higher commission on bulk purchase and also because the dealers supply at Re 1 less than the MRP. Shopkeepers are also lured by Sardar by using this tactics. Also sale tax is not paid on least 60% of sales and are unbilled/unrecorded by the retailers. Despite these, Sumul estimates that the present daily sale of Sardar is not more that 4000 litres and poses no threat to Sumul due to its fair market practices, quality standards, regular supplies and relatively higher commission (paise 33/per litre, with seasonal changes in commission linked to sales price of the milk). Sumul thus has 80% market share in Surat (D Shah, 1990).

Marketing strategies and trade practices

Our experience and impressions on private dairies' marketing operations are as followas :

(1) The main reason why retailers are attracted to private dairy is the high margin – while Sardar pays 50 paise/litre, co-operatives pay between 8-33 paise.

(2) Though they collect milk from the rural areas and transport it to Ahmedabad for processing, yet they can afford relatively higher procurement price, higher commission though selling milk almost at same price as the co-operatives on account of unfair trade practice as well as use of the available net work.

(3) Private dairies do not make any effort to create market but use the co-operatives' retain infrastructure and thus limit overheads.

(4) Low overhead cost is also due to the limited range of market they cover, many catering loose milk to a small clientele through home delivery within a limited area .

(5) In Ahmedabad, 88% of the retailers sell Amul milk due to consumers' preference, while 15% sell Sardar. Private dairy product sales is due to relatively high margin and credit facilities (in Ahmedabad 58% of the private dealers were getting one day to one week's credit).

(6) Unlike the private dairies, most of the co-operatives follow a policy of replacement of leaked/damaged pouch only at the point of unloading (it is difficult for a retailer to check every crate/pouch at that point of time and so he has to bear the loss).

(7) It is a standard practice of the retailers to push private brand due to higher margin. Most of them, after 8 am, attempt to push the private milk in lieu of co-operative brand on the plea that latter is unavailable.

(8) At times, private dairy/traders are able to create an impression that the private milk is better as no powder is used, cow milk is not added in buffalo milk and that fat is more than what is stated.

(9) The standard practice of private dairies is to sell the milk to agents/retailers at less than maximum retain price (MRP) which adds to retailers' margin. The difference between the MRP and actual sales price ranges between Re 0.25 in Ahmedabad to Re 1.00 in Surat.

(10) Evasion of 14% sales tax is also common (60% sales with and 40% without tax).

(11) Violation of standards of weights and measures is believed to be common – milk is not available according to printed fat and SNF though such cases are not established.

(12) Adulteration by water and high % age of bacteria etc are quiet known in the case of loose milk. Leaked packets are repacked with substandard milk and sold.

Suggested policy measures to promote co-operative milk

Though co-operative marketing strategy is designed by the Federation, Unions have adequate freedom to formulate their own liquid milk marketing policies. We may spell some important considerations in their review of liquid milk marketing strategy.

(1) Through a number of ways the demand for higher margin to the distributors could be met, namely : relate margins to sales price, have a system of graduated margin by relating margins to number of crates sold, incentive margin for locating new buyers and margin for exclusive sale of co-operative brand. Moreover, some kind of uniformity in margins among the Unions also necessary (present variation between 18-33 paise is too high).

(2) Develop an active network of auto-vending centres in different corners of the city to offer milk from 6 am to 10 pm.

(3) Consider a common co-operative brand to consolidate the overall co-operative share and to exploit the favourable image of the co-operative among consumers in terms of quality.

(4) Intensify consumer education on aspects of quality etc as 35% consumers do not have any idea about what quality of milk they are buying (as shown in the Ahmedabad study) at all and are unfamiliar with terms like fat, SNF, Standard milk, whole milk etc.

(5) Reduce distribution cost to achieve lower overheads (we understand that the Federation has already initiated steps to this end).
(6) Minimise 'leakages' and adopt liberal replacement.
(7) Explore possibility of home delivery system.
(8) Expand the market through effective campaign.

FUTURE IMPLICATIONS OF PRESENT DEVELOPMENT – POLICY IMPERATIVES

On the basis of the available evidence in Gujarat in general and Mehsana milkshed in particular, this paper attempts to examine validity of the arguments in favour of liberalisation especially those of the World Bank and to analyse the implications of liberalisation on the co-operative dairy industry of Gujarat and suggest broad policy actions. Before we do so, let us briefly recapitulate the Mehsana field experience :

> Major private dairies are as many as 35-40 (most prominent being Sardar), accounting for a total procurement share of some 20-25%, severely affecting co-operative procurement in 1994-95. Private efforts are more systematic and organized than ever before, posing serious threat to co-operative organisations. They concentrate on villages which are close by and have high density of milk surplus, to the disadvantage of the distant and low surplus areas. They mimic co-operatives to create an image of effective alternative which would provide better price, market and support in inputs, credit etc. This has weakened the faith and loyalty of members producers to some extent. They have become successful in increasing their market share in liquid milk markets and have slowly changed the structure of the market from a sellers market to a buyers market.

Invalidity of the World Bank's pro-libealisation arguments

(1) World Bank contention of co-operative monopoly is far from reality, as more than 80% of dairy business is open and available to private sector and already 35-40 units of varying capacities are actively operating in Mehsana and applied for MMPO registration. However, they could not get clearance on valid grounds. In milk marketing, their presence is easily observed.
(2) The argument of co-operative "inefficiency" in term of lower procurement price, higher processing cost and consequent higher consumer price is untenable, as illustrated by an earlier study of our own as also by a recent comparative study (Parthasarthy, 1995).

(3) Argument that high investment cost would have to be incurred in developing new milksheds by new private firms is unreasonable. No private dairy including Sardar have shown any inclination of making any investment in production enhancement but tend to encroach in co-operative villages as their strategy is to minimise cost and maximise profits and not to benefit the producers. The developed milksheds offer opportunity to them to derive free benefits of co-operative investment which is unjustified.

(4) That the World Bank considers the stand taken by co-operatives against state subsidies to private sector as a double standards is a clear evidence that it has totally failed to appreciate socio-economic environment of Indian dairying, role of co-operatives and our national policy. To quote LC Jain ("Cooperatives : Waiting for reforms", India Today), "Co-operatives have much more for the poor than the State-run anti-poverty schemes". During the second phase of Operation Flood, around Rs 273 crores were spent which brought Rs 3.66 million marginal farmers within the ambit of the milk co-operatives – in sharp contrast, the Government's IRDP, during the same period, spent Rs 1700 crores which offered benefit to only 2.5 million families. Do World Bank expect the private sector to assume this role in social development?

(5) Finally, the World Bank opposes the MMPO barriers to private entries, on ground of sale of contaminated/adulterated products, as a wrong approach to resolve the problem and suggests to streamline the regulatory and enforcement agencies. The World Bank officials should see for themselves the working conditions of private dairies applying for MMPO registration in a relatively advanced state like Gujarat. Even after MMPO, the existing laws have failed to contain adulteration etc and adulteration has become far more complex and more injurious.

Suggested policy interventions

The policy of dairy delicensing is now a reality with little possibility of policy reversal. On the contrary, it is quite possible that even the present "toothless" revised MMPO may be withdrawn as the central and state governments do not really appreciate any effective role of MMPO :

> "It is reported that Indian Government has directly given MMPO registration to Anmol dairy of Mehsana district and also permitted it to collect milk from 7 districts of Gujarat ... 20 out of 56 applicants have been granted registration ... two units were granted

permission, overriding the recommendations of Evaluation Committee.." (GCMMF News Bulletin, Dec'96).

Right now, the major threat to co-operatives is not from big private companies, but mushrooming small units who are capable of avoiding regulations and laws, who adopt all means to destabilise the co-operative unit. Once the second wave of radical liberalization sets in, high investment from domestic private companies, collaborators and MNCs are most likely to jump the band wagon. It is quite possible that some of the existing private units would be taken over by MNCs in future in which case, not only liquid milk but also the product market also will witness wild competition. It is almost certain that co-operatives which do not work efficiently, will not be able to survive.

(1) Suggested actions at village society level

One of the strategies the co-operatives could follow is exemplified by the milk co-operative in Charada village (Vijapur, Mehsana) where per family annual dairy income has attained height of Rs 16000 and, as a result, producers say with confidence and pride that there is no question of private dairy at all ! It is this economic gain which has generated strong loyalty to the co-operative. Another related area is promotion of production and productivity to meet the increasing producers' expectations vis a vis private procurement price as higher productivity and lower cost of production would offset the threat posed by artificial private price. All efforts must be made to promote such objectives. On-going efforts in co-operative education should be intensified, targeting the weaker sections and the women, and special training should be offered to those who demonstrate dynamic leadership qualities. Honest business practice, honourable and impartial treatment and prompt grievance redressal system should be enforced to imbibe the sense of collective ownership and responsibility at the society level which would minimise the need to adopt restraining measures.

(2) Suggested actions at the Union level

At the district Union level, the main objective should be to achieve all-round efficiency. Specifically, the following aspects deserves attention:

(1) The processing margin, believed to be much higher than international standards, must be contained to be able to offer better price to producers. Technical losses be reduced to minimum acceptable level (we gather that the average handling loss has already been reduced from 1.8% to 0.5%). Similarly, fat and SNF

loss be reduced to minimum.

(2) To future capacity expansions, internally owned capital must be expanded so that it becomes self-reliant.

(3) (The procurement price policy must be based on some rationale – even if the cost of production is not possible to cover fully, the out of pocket expenses and the increasing cost of living of rural producers given due consideration.)

(4) Because of combined effect of better marketing of milk (forward integration) and better supply of inputs/services (backward integration), Gujarat unions are able to generate considerable surplus, compared to Maharashtra where apparently farmers are receiving higher procurement price. (Pande, Mathur, Rang Rajan and Gupta, 1985). (Hence, effort should be made for production and supply of inputs and extension at least cost to promote productivity.)

(5) Finally, the manpower policy needs review. It is high time that steps are taken to retain skilled personnel and technicians through a combination of promotion, incentives and human resources development. It should be realised that the midwives of private dairies in Gujarat are, in most cases, none other than the former employees of co-operative dairies, a trend which could be curbed only by adopting a positive labour management.

(3) Suggested actions at the Federation level

Federation is the focal point in designing policies and strategies to safeguard the interest of co-operative milk unions against private competition. Competition as such is not new to Gujarat co-operatives as Anand itself was formed as an antidote to the monopolistic Polson. What is required is for co-operatives is to rediscover their strength and recognise the weaknesses and adjust themselves to the changed situations. A few policy suggestions are as follow:

(1) Make all out efforts to maximise the milk market share so that the private dairies are reduced to a level of insignificance.

(2) In the new circumstances, the conflict of interest among co-operatives is likely to increase and hence the Federation will have to play a rational, positive, objective and unbiased role to maintain its position among the unions. Some such areas are : market share in major cities, procurement sales price, milk sales obligation, sales price, inter-union milk transfer, allocation of product manufacturing, milksheds sharing etc. Needless to say that the stronger the Federation greater would be its capacity to ward-off private competition.

(3) The Federation must take the lead in persuading the Government

to initiate the following institutional, legal and fiscal changes :

- Formulation of a new Cooperative Law of Gujarat, on line of Model Act of the Planning Commission.
- Rationalising the Tax policy for co-operatives.
- Enactment of Cooperative Company Act.
- Appropriate Labour Laws for co-operative dairies.
- Effective implementation of MMPO.
- Restoration of the policy of export/ import canalization through NDDB (Shah, 1996).

REFERENCES

1. AH Somjee and Geeta Somjee (1978), "Cooperative Dairying and the profile of social change", Economic Development and Cultural Change, 1978.
2. CG Ranade, Mathur, Gupta and Rangrajan, "Performance of Integrated milk co-operatives — A study of selected co-operative dairies in Gujarat and Maharashtra", CMA, 1984.
3. JP Jain and BC Saxena, "Impact of milk supply schemes on the Rural Economy in milk collection areas of Dudhsagar Dairy, Mehsana", ICAR, New Delhi, 1977.
4. JJ Baxi and Sumindra Battacharya, "Strategic Business issues in Dairy co-operatives of Gujarat" Symposium on Management of Rural Cooperatives, IRMA, December, 1992.
5. Ministry of Agriculture, Department of Animal Husbandry and Dairying, MMPO-1992, 9th June 1992, New Delhi (1995).
6. Ministry of Agriculture, "Interim Report of National Commission on Commission on Agriculture on milk production through small and marginal farmers", 1971.
7. Parthsarthy S, "Institutional choice in Dairy Development — A comparative analysis of co-operative, government and private sector forms of organisation", 1992.
8. Planning Commission, Sixth Five Year Plan, 1980.
9. PC Bardhan, "MMPO, Steps in Right Direction", NDDB Dairy, May 1993.
10. SM Patel, DS Thakur and MK Pandey, "Impact of milk co-operatives in Gujarat", Institute of Cooperative Management, 1975.
11. Tushar Shah :
    -"Institutional Structures for Dairy Development India's Post Independence Experience", Indian Dairyman, January 1995.
    -"Liberalisation and Indian Agriculture — New Relevance of

Farmer Cooperative", Key paper in 55th Annual Conference of Indian Society of Agricultural Economics, Indian Journal of Agriculture Economics Vol. L July-Sept 1995, No 3, pp 488-507.

12. United News of India (1993), Dairy Industry - Deregulation for local exploitation, Economic Times, Calcutta.
13. United States Internal Society, Comparative, Study of Public Sector and Cooperative dairying, Economic Times, 6th May 1993.
14. VS Vyas and KM Chaudhari, "Economics of Dairying, Farming in Mehsana District", Arthvikas, July, 1971.
15. World Bank, "India, Livestock Sector, key issues for Emerging Growth and Development", May 25, 1995 (unpublished).
16. Manish Gupta and Sunil Kumar MA Comparative Study of the liquid Milk Marketing : Cooperative Versus Private with respect Ahmedbad, 1996 (unpublished).
17. Gujarat Cooperative Milk Marketing Federation Ltd., "Gujarat Milk Union - An Overview", February 1996 (unpublished).
18. Dilip Shah :
    - "An Economic Analysis of Cooperative Dairy Industry in Gujarat", Somaiya, Bombay, 1986.
    - "Cooperative Development in India - A need for Fresh Approach, Initiative and Action" (unpublished Monograph), 1995.
    - "Cooperative legislation in the context of economic liberalisation in co-operatives and economic development" edited by Dass, 1996 (Forthcoming).
    - "Cooperative Reforms in Gujarat", Financial Express, 1995.
    - "Dairy Cooperation — A Instrument of Rural change", Rawat Publishers, 1992.
    - "Role and Future of Cooperative Movement in an Environment of Economic Liberalisation in co-operatives and economic development: (edited by Dass), 1996 (forthcoming).
    - "The Gujarat Cooperative Milk Marketing Federation — An Appraisal of its functioning and organisation" Dr. Dilip Shah (edited) in Cooperative Milk Marketing and pricing practices of Gujarat, 1990. South Gujarat University, Surat
    - "Delicensing Dairy Product manufacturing, Logic and Reality", Financial Express, May 9-10, 1992.
    - "A brief Report of General Evaluation of District Union — A case of Mehsana District Cooperative Milk Producers Union, Mehsana, Gujarat" (1978-82).
19. Centre for Monitoring Indian Economy (CMIE), Profiles of Districts, 1993.

20. Directorate of Animal Husbandry, Gujarat State, Bulletin of Animal Husbandry and Dairying Statistics, 1992-93.
21. Gujarat Cooperative Milk Marketing Federation Ltd., Annual Reports : 1992-93 to 1994-95.
22. Mehsana District Cooperative Milk Producers Union Ltd., Annual Reports 1991-92 to 1994-95.
23. National Dairy Development Board, Annual Reports 1993-94 and 1994-95.

□□□

8

Private Sector Dairies in Vadodara : Impact and Future Strategy

1. Introduction

This micro-level study follows the earlier macro-level analysis of delicensing of dairy product manufacturing (Shah 1992). The first such micro-study was that of the milk union of Mehsana, Gujarat's richest milkshed (Shah 1996a). The Vadodara District Co-operative Unions' (Vadodara Union) dairy is located in an urban area where the liquid milk market dominates. Thus, this study will give a more complete picture of liberalisation in the dairy sector. This would help to develop a sound strategy for co-operatives to meet the changes in their business environment and relations with the State successfully.

2. Review of Dairy Policy

The National Commission on Agriculture had recommended, after studying the Anand Pattern, development of co-operative dairying through small and marginal producers and the landless (Ministry of Agriculture 1971). Operation Flood, an instrument for replication of the Anand pattern, was given positive recognition in the five year plans (Planning Commission 1974). A policy reversal in July 1991 delicensed the dairy processing sector and exempted this industry from the Industrial Development and Regulation Act, 1951. However, within a short period, the Milk and Milk Products Order (MMPO) was promulgated (Ministry of Agriculture, 1992). The provisions of MMPO were to regulate the process of liberalisation in such a way that the interest of consumers, producers, co-operatives and even the government could be protected.

The private sector and World Bank severely criticised MMPO as "backdoor licensing" (World Bank 1995). The World Bank argued that MMPO will result in local monopolies in a milkshed, high processing margin because of restricted competition, higher consumer price, and discourage producers.

The private sector succeeded in tilting the policy in its favour. In May 1993, major provisions of MMPO were so diluted that the modified MMPO became a "toothless tiger". It has opened the gates for the mushrooming of tiny units in the dairy sector who are not "amenable to regulation" and "exposed" to public gaze (Shah 1995). This is regarded as a major threat to the co-operative sector, more difficult than the competition with MNCs and major companies.

Though systematic data on the status under MMPO for the whole country are not available, information from informal sources are given in Table 1. Expectedly, bulk of the applicants (63%) were from the private sector. While 58% of the existing capacity is in the co-operative sector (including government dairies), its share in incremental capacity is only 14%, should all the applicants be granted license. Only one-third of the eventual total capacity would be with the co-operatives and, more ominously, the country's total capacity would be almost twice the total rural marketable surplus. In addition, milk potential and institutional coverage vary across states. Table 2 gives state-wise number of applications under MMPO. The highest number of applications came from Uttar Pradesh (which had 18% of India's milk production in 1994) and more than 50% of the applications had come from north India (which had 45% share in production). In Gujarat, the stronghold of milk co-operatives, similar developments are taking place, as per press reports.

3. Study Design

This study focuses on the following questions :

What is the milk production, marketable surplus, installed capacity and new capacity, either applied for or approved, in the Vadodara milkshed ?

What is the share of private dairy operators in total milk procurement ? What are the factors responsible ?

What is the share of private dairy units in the Vadodara milk market ? What are the factors responsible ?

What are the implications for the Vadodara Union ? What are its strengths and weaknesses ? Can the union compete ?

What should be the strategy and policy actions of the union ?

3.1 Methodology

Apart from using secondary sources of data, we visited the dairies as well as village co-operatives and met several groups and individuals. At the Vadodara Union, we met the managing director and the managers of procurement, marketing, co-operative development etc. We held discussions with the field level supervisor of the co-operative development project (CDP) and a few marketing supervisors. Another important source of data was the comparative study of one of the most developed village dairy societies, Dabhasha with the most underdeveloped society at Chitral. We studied how the private dairy operators could be successful in the Padra taluka of Vadodara district. We chose Padra because it is not only close to the Vadodara dairy (22 km) but also one with the largest milk production and procurement, attracting private operators. In fact, the first private dairy has been set up in Jasapura village of padra taluka.

The earlier study in Mehsana provided comparative data. We also discussed the issue of private dairy operations with the erstwhile city based milk vendors and milk product manufacturing units.

4. Profile of the Vadodara Milkshed

Vadodara district is located in the centre of Gujarat, with an area of 7794 sq. km. Its population of 30.4 lakh is dominated by Rajputs and tribals. The CMIE relative index of development of 160 is higher than the state average of 114 (CMIE 1993). However, some talukas are drought prone. The forest cover is 10.58% of the total area, marginally higher than the state average of 10%. The proportion of irrigated area is 20.92%, just close to the state average of 22%. The main sources of irrigation are wells and lift irrigation.

The average size of holding is 2.57 ha, lower than the state average of 3.29 ha. The main crops are cotton (37% of gross cropped area), arhar (18.53%), rice (10.64%) and jowar (10.5%). Other crops are tobacco (5.22%), maize (6.33%) and bajra (65%) (District Handbook of Census, 1991).

The mileh animal stock and changes in its composition during 1982-92 are given in Table 3. Between 1987-88 and 1992-93, unlike in other areas, stock of buffaloes declined by 13% while that of local cows increased by 24%. Crossbred cow stock also declined by 29%. Between 1982-83 and 1992-93 the proportion of buffaloes and cows marginally changed from 69 to 66% and 31 to 34% respectively.

Table 4 gives a comparative picture of the Vadodara milkshed in Gujarat, as of 1992-93. Vadodara ranked ninth in the state. The average yields per day of cows and buffaloes were 1.38 and 3.07 litres respectively, lower than the state average of 2.35 and 3.58 litres.

The relatively large average landholding, dominance of cash crops like cotton and tobacco, and relatively greater employment opportunities in and around Vadodara, do not make dairying as popular as in Kheda and Mehsana. Dairy income among Vadodara milk farmers is also relatively less significant.

4.1 Marketable Surplus

Marketable surplus determines procurement, capacity utilisation, and thus the efficiency of the dairy plant. Available data on yield per animal and stock of milch animals, from the Directorate of Animal Husbandry, and the CSO estimate of rural livestock and urban consumption, permit some broad estimates (see Table 5). It appears that even the existing capacity is more than net marketable surplus. The Vadodara dairy in fact faces a problem in meeting the urban demand.

According to some informal sources, the Vadodara milkshed has attracted five applications from the private sector to install new capacity of 4.50 LLPD. If granted, the estimated deficit would be as high as 5.84 LLPD.

5. Private Dairy Operators' Share in Procurement

The entry of private dairies is through procuring small quantities from good milksheds, natural or developed by the Vadodara Union. In 1995-96, some private dairies like Madhav, Bumia, and Ellora Park could increase their procurement as shown in Table 6.

The private operators also have some kind of an organisation called "customers' association". Consumers sometimes confuse them with co-operative societies. The private milk collectors use non-standard measures (khangi map). They also accept milk on kilogram fat basis, though there are serious malpractices.

Table 7 shows the taluka-wise scenario. There are about 661 private persons in 12 talukas who collect around 37000 litres per day, just 7% of milk collected by the **Vadodara Union.** However, the private sector is concentrated only in those areas which are nearest to the market and best in milk production/procurement. While the private societies are located in Padra, underdeveloped tribal areas of Nasvadi, Chotta Udaipur, Jetpur Pavi, etc., have negligible presence of the private sector. Developing such areas is the major argument of co-operatives against delicensing. Between 1995 and 1996, there has been an increase in private dairy societies and their procurement by 42.18% and 9.63% respectively. Thus, the increase in procurement has been much less. To sum up, the private dairy operators have successfully entered the milkshed.

5.1 Factors Favouring Private Dairy Operators

We studied two village co-operatives in Padra taluka which is the most affected by private sector entry, and also has a major private dairy. The two societies are highly contrasting cases and provide interesting clues (see Appendix). These factors have been identified through our discussions in the field.

In Vadodara milkshed, the credit-milk linkage, in a variety of forms, is the most dominating factor. Interest-free credit to buy milck animals is available. Some producers are also given input credit. Repayment in cash is generally discouraged, but insisted through milk supply. Another credit-milk linkage developed by private traders is Integrated Rural Development Programme (IRDP) finance. Such a linkage was absent in Mehsana. These private operators have strong linkages with District Rural Development Agency (DRDA), banks and insurance officers, and help producers to secure such finance and subsidies. As a result, even non-poor are able to secure the benefits of IRDP.

Another important factor is the price policy and mode of payment. The private dairy operators pay Rs. 10-15 per kg fat more than the Vadodara Union. Moreover, most pay on the same day. If the supply is regular and reliable, one could get advance money also. These are effective particularly when the producers face cash shortages for social expenditure.

The private operators also attract producers unable to meet the impersonal discipline of co-operatives such as regular supply, specific arrangements for collection and fixed time schedule. Large producers supply to private operators who come to their doorstep.

There is a false perception of better price being paid by private vendors for low fat or even adulterated milk. Some producers sell milk to both private traders and co-operative. This has the dual advantage of vendor's price and the subsidized inputs from the co-opeative. Such members get away because of political patronage.

Adulteration of milk is a much more prevalent practice in Vadodara, compared to Kheda and Mehsana. We were told that it is a custom that unadulterated milk should not be supplied at all ('Kumaru dudh no vechai'). There is a demand from the producers to abolish solids not-fat (SNF) measurement by the union and payment on the basis of fat alone. It was reported that there are societies which pour milk with SNF below standard and these SNF losses have to be borne by the union. The private operators accept adulterated milk and brand it as cow milk. Widespread corruption in the government inspection machinery enables private vendors to sell adulterated milk as cow milk.

"Early milk", a low fat milk produced just after the birth of a calf, is not accepted by the Vadodara Union. The private dairies accept it. This temporary shift many a time becomes a permanent one.

The conventional contractor system of collecting milk is still retained by the private traders. However, they have developed a semi-formal system of societies and fixed agents, such as "customers' associations". The private sector is now mimicking the co-operatives by providing symbolic assistance in the form of feed and paying lip service to production enhancement.

It is believed by many that private sector is growing, not because of its efficiency but because of the inefficiency of the co-operatives. Some producers were not loyal simply because of irregularities, dishonesty, disregard, discrimination, and indifference to their complaints.

Finally, socio-political factors have also had their effect. The Vadodara district is a hotbed of politics. Rivalries between the Congress and the BJP, and between Kshatriyas and Patidars have contributed to the prevailing situation.

5.2 Response of the Vadodara Union to Competition

A quick review of the annual reports of the Vadodara Union betrays the perception that, while there are potential threats in the milkshed, the real threat is in the milk market. As a result, the union has taken almost no steps in procurment. While the reports frequently mention the importance of quality and an effective distribution systems convenient to customers, they spell out no specific measures to achieve these. While recording the rise in milk sales, the reports are silent on the share of private dairies in the Vadodara milk market.

6. The Private Dairies in the Vadodara Milk Market

The Vadodara Union has a significant presence in the liquid milk market. Except for 20,000 litres of milk per day converted into products, as much as 80% of its milk is sold as liquid milk in Vadodara city. Hence, the management is more sensitive to the market share of private operators in the liquid milk market.

A recent study (Gupta and Kumar 1996) made the following observations on the problems arising in the Ahmedabad milk market following liberalisation:

The share of the co-operative sector declined from 70% in 1991-92 to 62% by 1994-95.

The market share of Amul is increasing but that of other co-operative brands like Uttam and Abad is decreasing.

The number of private brands has now gone upto 32 from 23. However, there are a few dominating brands.

There are at least five types of milk available to the customers.

6.1 Estimated Market Share of Private Operators

The city milk sellers estimate that around 90,000 to 1,00, 000 litres of milk per day are collected and used by about 200-250 small and big dairy operators/traders. Most of the milk bought by them is used for making milk products. Liquid milk is sold depending on availability. However, fixed daily customers are served regularly. Milk of all kind is taken as 'cow milk'. The price is always Rs. 15 per kg fat more than the union. SNF is not considered in general. Milk is sometimes accepted on value basis. There are advances and periodic payments. However, accounts are settled annually during Diwali. Except this, no other services are provided. There is no binding on any party.

The growth of private sector in the liquid milk market has been estimated by several sources. The Gujarat Co-operative Milk Marketing Federation (GCMMF) has been monitoring the milk market in the state since 1991. The 1996 report is in Table 9. The share of the major private dairy in Gujarat, i.e. Sardar Dairy is around 50,000 out of the 2,75, 000 litres sold by the private sector brands (see Table 10). Since the last two years, the Vadodara dairy surveys the market through their field officers (Table 11).

The market share of the Vadodara dairy which was above 70% in 1989-90 has declined to 60% in 1994-95. The number of dairy brands increased, indicating tough competition. The share of private brands has doubled from 7% to 15%. However, this rise is at the cost of vendors. The local but organised private dairy unit (LMP) is not yet a significant operator.

6.2 Practices of Private Trade

The private dairy operators collect milk from the rural areas around the city, as they need a relatively small quantity. The retailers are attracted to private dairies by the high margin: the Sardar dairy pays 50 paise per litre; the Vadodara dairy 14 paise; and Vasudhara (LMP) 50 or even 60 paise. However, LMP's operations are constrained by procurement and hence it is unable to play a significant role. Private operators have a significant advantage of a milkmarket developed by the Vadodara Union. As they do not create their own retain infrastructure, they have limited overheads. The private dairies give credit from one day to one week to retailers. The Vadodara dairy follows the policy of replacement of leaked or damaged pouches only at the point of unloading. It is difficult for every retailer to check every crate and pouch. Buyers have strong preference for the co-operative brand. However, retailers push private brands because of higher margin, especially after 8 a.m. on the plea that the co-operative brand is not available. At times, the private dairies/traders are able to create an impression that private milk is better because the dairy milk is not fresh, not creamy and added with milk powder. They sell milk to agents/retailers at less than the maximum retail price (MRP) which adds to the retailer's margin. Evasion of sales tax is a common practice as sale takes place without a bill. Violation of standards of weight and measures are believed to be common. Adulteration is quite common.

7. Performance of the Union

The private operators have a sizable share in the Vadodara market. However, only smaller units are now operating, depending on outside milksheds for their procurement. In the long run, with the entry of larger private business, the pressure on procurement will increase substantially.

Therefore, we should design an appropriate strategy. We discuss briefly the following broad indicators of performance : institutional development ; procurement; marketing; business performance; production enhancement; and social and co-operative development.

7.1 Institutional Development

Institutional development is a key to success. The patronage of the members depends upon the efficiency with which the Vadodara Union can protect and enhance the economic gains. This, in turn, depends upon producer-oriented institutional development. Specific indicators of institutional development during 1991-95 are given in Table 12. Table 13 provides a detailed picture of the spread and coverage of the institution at the taluka level.

The union has covered only 51.67% of the villages. The coverage is quite uneven; in Padra and Savali, 93 to 95% ; in the Chota Udaipur tribal area, only 7%. This may be because of distance and production potentiality of the region. However, the growth rate of the societies has remained less than 1% except in 1992-93. In 1994-95, 15% of the societies were closed. This closure of the societies in the tribal area is not necessarily because of private dairy operators but also because of non-availability of surplus milk and absence of managerial and organisational support to continue the operation in difficult times. In the tribal taluka of Chota Udaipur, the percentage of closed societies was as high as 53. There are unions where the percentage of functioning societies is as high as 99. Only in Bharuch (40%) and Rajkot (53%) the percentage of closed societies is higher than that in Vadodara.

The annual growth rate of members is similarly less than 1%. Experts suggest at least 200-250 members for viable operations. The Vadodara Union is falling quite short of this. Viability is of course also determine by quantity of milk collected. The average procurement per society has remained in the range of 149-171 kg per day, against a norm of 500 litres, or 515 kg (FAO 1972). The co-operative department itself suggests 100 litres or 103 kg per day to achieve economic viability (Ministry of Co-operation 1976).

It is now widely established by National Sample Survey (NSS) studies that dairying is an occupation of rural women. Hence modernisation could be accelerated if women become active participants. Though the Vadodara Union has no exclusive women dairy co-operatives, upto 20% of its members are women compared to Mehsana where the figure is as much as 33%.

7.2 Milk Procurement

The indicators are given in Table 14. Procurement as a percentage of total production shows a significant decline : 39 to 24. In three out of five years, the percentage rise in milk procurement price was higher than that in the agricultural labour price index. Despite this, procurement has declined in all the years except 1993-94. It has gone below even the procurement level of 1989-90. As a result of the procurement pressure, the price and cost of procurement have gone up substantially. For example, the per litre cost of transport has gone up from 19 paise to 27 paise between 1991 and 1995, not all of which is due to the increase in fuel costs. Another consequence was purchasing milk from other dairies; sometimes this has been as much as 50% of throughput

Seasonal fluctuations have reportedly increased even in the relatively developed milksheds. Cow milk production was providing some relief in maintaining stability. However, the proportion of cow milk has also gone down.

7.3 Business and Financial Performance

Overall, sales have gone up by 17% during the period 1991-95 (Table 15). However, the growth in sales is not matched by procurement growth.

As a result, the deficit was met by purchasing milk from other dairies. Such dependence has increased from 1.97% to 26.80% of sales. Despite this shortage, the Vadodara Union has continued with product diversification. However, the growth rates of products like curd, ghee, butter, flavoured milk, etc., were either negative or only marginally positive (Table 16). The Sugam brand was continued but growth rates were negative except during 1994-96. However, during the period 1991-95, sales of *shrikhand* grew by 393%. Sugam products contribute around 18% to 19% of revenue (Table 16). The Union has developed an excellent distribution system with 700 outlets.

7.4 Financial Performance

The performance of the dairy co-operative cannot be judged from annual financial statements alone. The share of farmers in total sales realisation has declined from as high as 73.81% to 43% because a significant percentage of milk is purchased from other dairies to maintain the market share. This also indicates increased cost of managing the dairy.

7.5 Production Enhancement Programme

Provision of technical inputs is one of the unique features of the Anand pattern. The union implements standard programmes such as artificial insemination, health cover, and balanced feed distribution. It is also providing services such as provision of exotic bulls for natural service, urea treatment programme, crossbred male calf rearing project, and progeny testing programme.

The growth rate in the number of artificial insemination cases is very high but erratic. However, the quantum per society is low. There is no consistent progress in the technical input programme, but the per litre cost has increased by 60% during the period 1991-94.

7.6 Social and Co-operative Development

Like other unions, the Vadodara Union implements social and co-operative development programmes which are liberally assisted by NDDB. The union encourages village societies to participate in developmental activities. Smokeless *chulla*, biogas plants, *Kisan van yojana*, dairy road project, construction of milk collection centres, group insurance scheme for crossbred animals, education visit grants and publication of news bulletin, etc., are some of the programmes. A systematic co-operative development effort focusing on four distinct elements of the co-operative structure, namely, membership, leadership, professional management , and social and political environment is being implemented. So far, 13,037 women; 12,728 men; 1006 managing committee members and 75 secretaries have participated in this programme. A *mahila jagruti* programme was organised in which 670 women participated.

8. Strategic Policy Interventions

Delicensing is now a reality, with little possibility of a policy reversal. On the contrary, MMPO may even be withdrawn. The major threat now is only from small units which are capable of avoiding regulation and laws (Shah 1995; Shah 1966b). However, once the second wave of radical liberalisation sets in, large investments from domestic companies, collaborators and MNCs are likely. Apart from this, one must also consider the inefficiency in management and operation of the union in evolving a strategy.

8.1 Suggested Strategy at the Village Level

Procurement and patronage of the member producers must be increased alongwith cost effective production in the long run. As a first step, operational management and the union's image must be improved. Quite a few producers quit simply because of ill treatment. Honest business practices, impartial treatment and a prompt grievance redressal system should be in place. There is a need to replicate the Dhabhas type societies and discourage the Chitral type. The latter gives wrong signals. A package of incentives and disincentives must be developed to promote procurement. The input cost is increasing, with no impact on production and procurement. Hence, management may adopt some bold policy measures of discrimination in input supply and price in favour of those who supply milk. Economic return is the key to retaining patronage. Though higher procurement price is the crude indicator, other advantages such as input subsidies, stable price and market, etc., must be explained to those producers who are unaware of these. The cattle loan scheme and cheap credit scheme should be developed by the union if its resources permit. It should be able to link up with IRDP and other programmes.

Co-operative education can help stabilise patronage and promote participation. On-going efforts must be intensified. There is need to promote dairying in new areas and among new people. Tribal areas could be developed on the lines of Sumul in Surat. Efforts should be made to win over those village co-operatives which had supplied significant quantities earlier.

8.2 Strategy at the Union Level

Two-axis pricing (i.e., both for fat and SNF in milk) must be firmly implemented. The malpractices relating to SNF must be strictly handled. Milk production by crossbred cows should not be adversely affected. It is very important to develop strong producer organisations in areas of high milk production density, to check the private producers and to pressurise them to commit resources for production enhancement (Shah 1995). Since such support will not be forthcoming, they activirely lose out. (The societies should actively will not be forthcoming, they would lose out.) The societies seek the goodwill of Non-governmental Organisations (NGOs) who are alive to social issues.

The union should achieve efficiency in processing marketing and supplying technical inputs. The operational cost is quite high, around

Rs. 2.00 per litre. There is a need to reduce all kinds of technical losses. At present, even systematic records are not maintained. The union must reduce the gap between milk collection and sales.

The union could best survive by a larger market share. However, its capacity is limited. As a long term strategy, it should aim for a larger market share by upward revision of commissions, liberal policies on replacing leaking milk packs, compensation and special concession for exclusive distributors etc. The union should have plans for image building and consumer education.

Ultimate returns to farmers could be achived because of better marketing and/or better supply of inputs and services (Ranade *et al.* 1984). Hence the union must pay equal attention to strengthening the supply and delivery system for technical inputs.

The union needs a review of its manpower policy and a system of appropriate incentives and penalties to motivate the employees. Greater delegation of powers with accountability will lead to dynamism. If it is possible, a plan must be made for reducing surplus labour. However, at the same time, steps will be required to retain highly skilled and experienced personnel and technicians. It should be noted that the midwives of private dairies in Gujarat are, in most cases, none other than the former employees of co-operative dairies.

8.3 Strategic Actions at the Federation Level

Competition as such is not new to Gujarat co-operatives. However, they have to rediscover their strength. The union will have to depend on GCMMF to compete with the private trade. GCMMF must take the lead in persuading the government to initiate institutional, legal, and fiscal changes; enacting a new co-operative law for Gujarat on the lines of the model act of the Planning Commission; rationalising the tax policy for co-operatives; enactment of the Co-operative Company Act; labour laws for the co-operative dairy industry; effective implementation of MMPO; and restoration of the policy of export/import canalisation through NDDB (Shah 1996b).

Conflicts of interests among co-operatives are likely to increase and hence the GCMMF will have to play an objective and unbiased role in resolving them. Market shares in major cities, procurement price, liquid milk sale obligations, sale price, inter-union transfer prices, allocation of products for manufacturing, milkshed sharing etc., are some of the potential areas of conflict.

The private dairy operators have already succeeded in procuring milk and capturing some market share, though not alarming at present. In this scenario, there is no time to lose.

References

1. Baroda District Co-operative Milk Producers' Union (1995) *Annual Report 1995.*
2. Centre for Monitoring Indian Economy (CMIE) (1993) *District Profiles,* CMIE.
3. Directorate of State Population Census (1991) *District Handbook of Baroda District.*
4. Gupta, Manish and Sunil Kumar M. (1996) *A Comparative Study of the Liquid Milk Marketing :Co-operative versus Private in Ahmedabad,* OYP 1995 PWM Report, IRMA, Anand.
5. Ministry of Agriculture (1992) *Milk and Milk Products Order* 1992.
6. Ministry of Agriculture (1971) *Interim Report of the National Commission of Agriculture on Milk Production through Small and Marginal Farmers,* Ministry of Agriculture, Government of India, New Delhi.
7. Ministry of Co-operation (1976) "Co-operative Policy and Programme of Government of India", *Co-operative Dairy Conference Souvenir,* National Co-operative Development Corporation, New Delhi.
8. Planning Commission (1974) *Fifth Five Year Plan,* Government of India, New Delhi.
9. Ranade, C.G., D.P. Mathur, B. Rangarajan and V.K. Gupta (1984) *Performance of Integrated Milk Co-operative; A Study of Selected Co-operative Dairies in Gujarat and Maharashtra,* Indian Institute of Management, Ahmedabad.
10. Shah, Tushaar (1995) "Institutional Structure for Dairy Development : India's post Independence Experience ", *Indian Dairyman,* 47(I), January.
11. Shah, Dilip (1992) "Delicensing Dairy Products manufacturing : Logic and Reality", *Financial Express,* may 9-10.
12. Shah, Dilip (1996a) *Dairy Product Export Performance : Potential and Policy in Post Liberalisation Period in India* (unpublished).
13. Shah, Dilip (1996b) *Entry of Private Dairy Operators in Mehsana Milkshed of Gujarat - A Rapid Assessment,* (unpublished).
14. UNO/FAO (1972) *Second Interim Evaluation of World Food Programme (WFP) Assisted Project,* India, FAO, Rome.
15. World Bank (1995) India : *Livestock Sector : Key Issues for Emerging Growth and Development.* The World Bank, Washington, D.C., May 25.

Appendix 1
Contrasting Features of Chitral and Dabhas D.C.S.

Chitral D.C.S.	Dabhas D.C.S.
Organised by a Rajput family and still dominated by it, as board of directors are family members.	Organised by 218 members in 1965 against exploitation by private dairy operators.
Membership declined to just 5 from 40 in 1989.	Membership increased from 218 to 550 members.
Earlier, procurement was 400 litres per day; but only 20 litres now.	It started with 77 litres per day but now provides as much as 1800 litres to Vadodara dairy after meeting the local demand of 200-250 litres.
Chitral is 40 km from Vadodara and yet there are as many as 6 private dairy operators who collect as much as 600 litres per day	Though it is very close to Padra, there are no private dairy operators at all. Earlier this village was an important source of milk to private operators. In fact, one trader recently tried but failed to penetrate.
The Co-operative society provides no services of any kind to its producers. In fact, it is almost a nominal, non-functional structure.	It provides many services. It has an AI centre and received the award for the highest milk yielding animal in India. It also has a unique scheme of providing gifts out of sample milk fund to all according to the proportion of participation. It produces a bonus of 13% to 15% almost every year. It has also purchased an electronic milk tester. It has also contributed Rs. 30000 to build a balmandir, bustand, water facility, public washroom facilities etc.
There is no election in the real sense since many years. The chairperson is a woman, her husband union even for voting to	The chairman and board get elected unanimously. A Patidar has been chairman since long, but the board is changing every year.

elect the board of directors to be union).	All are given the opportunity by consensus and rotation rather than by election
This "dummy" co-operative has neither strength nor the desire to compete. After the commissioning of the narmada canal, this area will produce more milk, but the present co-operative has neither the imagination nor any initiative to take advantage of it.	Dabhas could compete with private trade.

Table 1: MMPO : Status as on June 95

	Co-operatives	Others	Total
Applicants(No.)	256	442'	422
Existing capacity (LLPD*)	253	169	404
Incremental capacity (LLPD)	73	465	581
Total capacity (LLPD)	308	634	942
Rural marketing surplus (LLPD)	-	-	456

*LLPD-Lakh Litres Per Day.
Source : Informal.

Table 2 : Region-wise Number of Applications (May'94)

North	
Delhi	22
Haryana	71
H.P.	5
J & K	3
Punjab	57
Rajasthan	41
U.P.	174
Sub total	373
South	
A.P.	42
Karnataka	23
Kerala	8

Pondicherry	1
T.N.	28
Sub total	102
East	
A & N Island	1
Assam	2
Central	
Bihar	10
M.P.	19
Orissa	10
Sikkim	1
Tripura	1
West Bengal	10
Sub total	54
West	
Daman & Diu	2
Goa	1
Gujarat	54
Maharashtra	108
Sub total	165
All India Total	694

Source : Informal.

Table 3 :Vadodara Milkshed Milch Animal Stock : Composition and Changes

	Buffaloes	Local cows	Cross-bred cows	Total milch animals	Ratio of buffaloes to cows
1982-83	142000	63000	980	205980	69.31
1987-88	185000	66000	9010	260010	71.29
	(30.28)	(4.76)	(810)	(26.23)	

1992-93	161500 (-12.70)	82000 (24.24)	6400 (-28.96)	243500 (-6.34)	66.34
Changes between 1982-92	(13.78)	(30.15)	(553.06)	(18.21)	

Figures in parauthesis percentage changes.
Source : Vadodara dairy documents.

Table 4 : Vadodara Milkshed :A Comparative Picture

	Animals inmilk to total population (%)	Buffaloes to total bovine population (%)	Share of milch animals in the (%)	Rank in mikh animal holding (No.)	Milk production share (1983) (%)	Milk Yield (kg/day) Cow	Milk Yield (kg/day) Buffalo	Milk production share (%)	Rank (No.)
Mehsana	47	62	9.50	01	13.35	2.87	4.37	13.51	0.4
Vadodara	36	40	4.92	09	4.59	1.38	3.07	4.80	10
Gujarat	40	39	100	-	100	2.35	3.58	-	-

Source : Bulletin of Animal Husbanday and Dairying Directorate of Animal Husbandry, Ahmedabad, 1992-93.

Table 5 : Estimate of Marketable Surplus

	No. of villages	Villages covered by co-operatives	Milk production (LLPD)	Rural surplus (LLPD)	Urban consumption (LLPD)	Net marketable surplus	Existing Processing Capacity (LLPD) Private	Co-operative/ government	Total
Mehsana	1119	1009	13.03	9.91	1.20	8.71	2.20	10.05	12.25
Vadodara	1651	920	4.66	1.71	2.89	-1.18	2.40	3.00	5.40
Gujarat	16776	11535	91.92	46.42	28.22	18.20	6.85	69.75	76.60

Source : 1. Bulletin of Animal Husbandry (1990-91)
2. Population Census (1990-91)
3. CSO Consumer Expenditure Survey (1990-91)

Table 6 : Milk Procurement by Private Dairies : Vadodara Dairy Estimate

	March 1995	March 1996	Change(%)
Ellora Park	11000	13000	18.8
Madhav	4000	8000	100.0
Bumia	3500	6000	71.2

Source : Vadodara District Milk Producers' Union

Table 7 : Private Milk Procurement in Different Talukas

Talukas	March 1996				March 1995		
	Co-operative Societies (Nos)	Private Societies (Nos)	No. of Producers pouring to privates (Nos.)	Daily procure-ment (Ltrs.)	Co-opera-tive Societies (Nos.)	Private Societies (Nos.)	Daily procure-ment (Ltrs.)
Padra	88	77	243	19240	86	70	19758
	(10.89)	(25.66)	(36.76)	(52.37)	(11.16)	(33.11)	(58.42)
Vadodara	61	43	107	6180	65	32	5023
	(7.54)	(14.33)	(16.18)	(16.82)	(8.44)	(15.16)	(14.99)
Savli	137	46	79	3865	139	31	3173
	(16.95)	(15.33)	(11.95)	(10.52)	(18.05)	(14.69)	(9.46)
Dabhoi	80	44	70	2352	75	22	2629
	(9.90)	(14.66)	(10.05)	(6.40)	(9.74)	(10.42)	(7.84)
Vaghodia	65	19	22	850	55	19	660
	(8.04)	(6.33)	(3.32)	(2.31)	(7.14)	(9.00)	(1.96)
Karjan	73	34	75	2191	. 66	18	1480
	(9.03)	(11.33)	(11.34)	(5.96)	8.57)	(8.53)	(4.41)
Sinor	37	19	28	919	36	11	532
	(4.50)	(6.33)	(4.23)	(2.50)	4.67)	(5.21)	(1.58)
Tilakwada	40	02	02	80	35	01	24
	(4.95)	(0.66)	(0.30)	(0.21)	(4.54)	(0.47)	(6.07)
Sankheda	113	13	28	700	109	07	412
	(13.98)	(4.33)	(4.23)	(1.90)	(14.15)	(3.33)	(1.22)
Naswadi	13	-	-	-	15	-	-
	(1.60)				(1.94)		
Jetpur-	80	03	07	360	74	-	-
Pavi	(9.90)	(1.0)	(1.05)	(0.97)	(9.61)		
Chotta-	21	-	-	-	32	-	-
Udaipur	(2.59)			(2.72)			
Total	808	300	661	36737	770	211	33508
	(100)	(100)	(100)	(100)	(100)	(100)	(100)

Note :1. Traders collect milk measuring and testing on fat.
2. Mavawalas are paying per hundred gram move equivalent.
3. Without testing (Lectometer), the private operators pay Rs. 8-10/litre.
4. Most of the traders give credit against milk supply in advance.
5. Figures in paranthesis show the percentage of the total in each column.

Source : Compiled and processed data collected from Vadodara dairy.

Table 8 : Taluka-wise DCS and Their Procurement Shares

	Milk Procurement (Kgs)	Share (%)	Average per Society	Co-operative Societies (Nos)	(%)
Padra	7016213	14.33	79730	88	10.89
Vadodara	3369421	6.87	52236	61	7.41
Savli	7585504	15.48	55338	135	16.95
Dabhoi	3881693	7.92	48521	80	9.90
Vaghodia	3644781	7.44	56073	65	8.04
Sinor	372346	7.60	100623	37	4.57
Karjan	873933	1.78	11972	73	9.00
Tilakwada	1400855	2.85	35021	40	4.95
Sankheda	4648228	12.90	55925	113	13.98
Naswadi	5652537	11.53	35556	13	1.60
Jetpur Pavi	4648228	9.48	58103	80	9.90
Chotta Udaipur	871100	1.77	41480	21	2.59
Total	47315539	100		806	100

Source : Compiled and processed data from Vadodara Dairy.

Table 9 : Private Trade in Milk Market

City	Sales by private trade in the city as a % of its sales in the state
Ahmedabad	37.98
Vadodara	36.49
Surat	7.29
Rajkot	18.24
Total	100.00

Source : GCMMF.

Table 10: Distribution of Sardar Dairy

City	Estimated share in total sale by Sardar (%)
Ahmedabad	50.25
Vadodara	40.20
Surat	8.04
Rajkot	1.51
Total	100.00

Source : GCMMF.

Table 11 : Annual Average Milk Marketed by Private Dairy Units in the Vadodara Market

	March 1995 Share (%)	March 1996 Share (%)
Vadodara dairy	61.90	60.31
Private	7.0	15.30
dairies	7.0	15.30
Vendors	31.10	24.39
Total	100.00	100.00

Source : Vadodara Dairy

Table 12 : Institutional Development : Union Level

Year	1991-92	1992-93	1993-94	1994-95	1995-96
No. of village Societies	894	929	936	952	951
Annual rate of growth of dairy co-operative societies(%)	0.67	3.91	-0.74	1.70	-0.10
Functional societies to total societies (%)	85	87	87	87	84
Annual rate of growth of producer members (%)	2.64	0.28	0.51	0.91	2.4
Average number of members	173	168	167	168	169

Annual rate of growth of women members(%)	12.0	7.41	00	3.30	3.22
Women in total members(%)	17.0	18.0	19.0	19.0	20.0
Average procurement per society per day (litres)	171	171	154	172	149

Source : Vadodara District Milk Producers' Co-operative Union.

Table 13 : Institutional Development : Taluka-wise

Taluka	Lean to flush ratio (%)	No. of village level dairy co-op. societies	No. of village (%)	Coverage of villages (Nos)	Closed soci-eties (%)	Closed soci-eties	Share in procu-rement (%)
Vadodara	36	62	112	55.35	14	22.58	6.93
Padra	50	88	95	92.63	10	11.36	14.61
Savli	37	135	136	99.26	18	13.33	16.61
Dabhoi	41	80	118	67.79	10	12.50	7.50
Vaghodia	54	64	95	67.36	14	21.87	7.99
Karjan	34	73	94	77.66	04	12.32	11.63
Sinor	34	37	40	92.50	03	8.10	7.66
Sankheda	42	112	219	51.54	23	20.53	13.10
Nasvadi	21	113	124	91.12	31	27.43	0.92
Tilakwada	21	40	117	34.18	06	15.0	2.88
Jetpur Pavi	39	84	213	39.43	11	13.09	9.57
Chotta Udaipur	55	21	279	7.52	11	52.38	1.79
Union	39	909	1759	51.67	160	35.20	100

Source : Vadodara District Co-operative Milk Producers' Union

Table 14 : Milk Procurement Performance

	1991-92	1992-93	1993-94	1994-95	1995-96
Milk procured vs. total milk production(%)	39.38	31.72	26.72	28.67	21.34
Growth rate in union's procurement(%)	-19.30	-6.69	12.69	-11.87	-5.46
Milk procurement per society (litres/day)	171.03	171.03	154.05	172.01	149.29

Rise in milk procurement price (%)	29.78	20.54	0.75	9.12	21.57
Rise in the price index of agricultural labour(%)	8.2	10.05	8.35	10.85	7.2
Per litre cost of procurement (paise)	5.94	6.62	6.77	7.16	12.43
Per litre cost for procurement transport (Rs.)	0.19	0.25	0.21	0.23	0.27

Source : Data collected and compiled from Vadodara District Co-operative Milk Producers' Union.

Table 15 : Collection and Sales of Vadodara Dairy

Reference year	1991-92	1992-93	1993-94	1994-95	1995-96
Milk Collection (cr. kgs)	5.60 (-23.91)	5.22 (-6.69)	5.89 (+12.69)	5.19 (-11.87)	4.90 (-5.46)
Milk sales (cr. litres)	5.71 (-9.64)	5.94 (+3.94)	6.04 (+1.80)	6.60 (+8.68)	6.70 (+1.99)
Purchased milk to total milk sold (%)	1.97	12.01	2.06	21.03	26.80

Figures in paranthesis are percentage changes during the year.
Source : Annual Report of Vadodara District Co-operative Milk Producers' Union (1995-96).

Table 16 : Business Performance

	1992-93	1993-94	1994-95	1995-96
Sales Growth rate (%)				
Curd	13.31	1.9	13.46	6.31
Ghee	-14.33	1.74	-14.30	-36.40
Butter	9.9	-5.9	-4.52	-
Flavoured milk	14.82	6.36	3.7	-6.05
Butter milk	57.83	58.14	-1.70	-

Sugam Products				
Ice-cream	-16.85	-5.21	-15.42	10.65
Shrikhand	-7.79	-0.79	9.38	16.53
Sugam products in total sales(%)	19.1	18.39	17.99	17.84
Growth of distribution outlets (%)	14.31	16.39	11.09	20
Increase in distribution cost per litre (%)	15.0	8.69	4.0	

Table 17 : Production Enhancement Performance

	1990-91	1991-92	1992-93	1993-94	1994-95	1995-96
Artificial Insemination						
Annual growth of AI cases (%)	10.03	-15	3.93	22.76	6.83	0.85
AI per society (Nos)	5.57	35.13	35.13	42.81	44.96	45.13
Success rate (%) of AI (Calf born)	12.89	12.55	12.96	11.14	16.75	14.30
Growth of cattle feed sales (%)	1.69	-17.76	-7.49	-5.29	15.91	-14.66
Per society cattle feed sales (M.T.)	34.58	28.25	25.15	23.64	26.95	23.01
Growth of sales of fodder seeds (%)	6.10	-25.0	-40.95	+70	144.63	-62.06
Fodder seeds distributed per society (kgs)	18	12	20	26	63	24
Growth in treatment visits (%)	-4.37	10.03	-0.59	3.93	22.76	0.85
Cattle treatment per society (Nos.)	170	256	146	173	163	166
Growth in Women participation						
Growth in women development seminars (%)		+ 148	-68	109	6.0	154
Seminar visitors per society (No)	04	10	03	06	15	

Source : Vadodara District Co-operative Milk Producers' Union.

9

Cooperative Dairy Development in 2000, A.D.

Introduction

Since hardly a decade distant remains to footstep in the 21st Century, it is quite thrilling to view about the development of India's fast developing dairy sector in 2000 A.D. This sub-sector of agricultural economy of India is most important for several reasons. Firstly, according to C.S.O. reports Milk and Milk products are second largest contributors to the gross output value with turnover of Rs. 10,000 crores annually next to 'rice' with a value of Rs. 18,000 crores. 'Wheat' remains as third with an output value of Rs. 71,000 crores. Thus, milk is India's second largest agricultural commodity. Secondly, the energy significance of the dairying sector could be seen from the fact that 67% of energy for the motive power in agriculture is also supplied by this sector. Thirdly, equal important human energy source is also drawn from this sector as 90% of the animal protein and 100% animal fat is derived by Indian vegetarians from the perfect food i.e. Milk. Fourthly, most important is its rural development significance as it has the capacity to generate employment and adequate income for the vast number of rural poor classes. It has been well documented fact that 70% to 75% households of small and marginal farmers and landless labourers who also are 'the target' groups of the anti poverty and rural development programmes are wedded to the dairing traditionally since ancient times.

Finally, dairing is a very significant sector to generate employment with lower unit cost of employment for the target groups of the rural India. For example, it is found that one rupee investment in the dairy sector could generate three rupees worth employment. It has also been established as better source of employment as compared to other alternatives. For example, an investment of 10 lakh rupees in dairying generate 290 person years of employment, while in the crop production same investment can generate only 120 person years of employment. In fact, in the best favourable environment dairying has found to provide as much as 78% of the total annual income to small and marginal farmers in some regions of Gujarat.

The future of Indian Dairy Development is very bright mainly due to the foundations created by THE OPERATION FLOOD PROJECTS I, II and III implemented since 1970. The project has helped transforming traditional dairying through network of 'Anand Pattern' dairy co-operative structure and modern dairy infrastructure through its large flow of investment of around Rs. 13,000 - 15,000 millions till today. "It is this remarkable development which draws our attention and inspires us to undertake an interesting exercise to extrapolate the evolving pattern and project the future might holding for us with great challenges and opportunities."

The present paper is designed to discuss the dairy development scenario in 2000 A.D. within the following framework.

I First part attempts to document India's targets of production and consumption of milk based on the current status of supply and demand of milk.

II Second part is devoted to discuss in what ways these targets will be attempted to be achieved.

III Third part raises the issues which would be emerging in the process of dairy development in 2000 A.D.

I

The most serious attempt for the projection of the milk production and consumption in 2000 A.D. was done by the National commission on Agriculture in 1976. It estimated 49-36 million tons of milk as low estimate and 64-40 million tons of milk at 1986-87 base as high level estimates of the milk production in 2000 A.D. This would result into the per capita availability of 125.38 grams in rural areas and 146.34 grams in urban areas on low side and 146.34 and 183.83 grams at high side of estimates in 2000 A.D. However, recently, the technology mission on dairy development has targeted the milk production at 2209 lakh M.T. milk in 1987-88 leading to per capita availability milk from 158 grams in 1986-87 to 196 grams by the year 2000 A.D. Thus, one may state that we plan to jump the production and consumption by 400% and 27% respectively from the level of 1987-88. This also implied the fact that by 2000 A.D. medically prescribed level of per capita consumption atleast in terms of availability we shall be able to achieve. This production projections are very much depending upon our capacity to achieve the projected yields of milk per animal in 2000 A.D. It is projected by the technology mission on dairy development that we shall achieve yield promotion by 108% from the level of 390 litres per annum to 800 litres per annum in case of cows and by % in case of buffaloes from the level of 910 litre per annum per animal in 1987 to 1100 litres per animal per annum in 2000 A.D. It may be noted that it is envisaged that we shall be able to raise cow milk yield by almost 5 times than current level yields. This implied that it will be required to have a great 'Technology push' to promote this much level of yield.

Though the official documents do not support the imbalance between the demand and supply, some independent study shows that the technology mission estimates of demand are rather under estimated. For example, it is projected that in accordance with the recommendation of Indian Council of Medical Research to ensure the per capita per day availability of 200 grams of milk total requirement of milk would be around 79.5 million tonnes for a human population of 990 million in 2000 A.D. To achieve this much production we will need to boost up production growth from the level of 4.6% per annum to 6.8% growth rate of production per annum by 2000 A.D. This level of production is likely to balance the demand which might be generated at the growth rate of 6% per annum assuming an annual rate of growth of 5% GNP as considered feasible in the earlier VIII plan approach; due to reduced growth of rate of GNP targeted by new VIII plan approach it would be more feasible to assume that there will not be sharp demand supply distortion. However, regional and seasonal gaps and gaps arising out of instable production required to be accounted to describe the realistic picture of the equilibrium of demand and supply of milk in 2000 A.D.

II

It would be interesting and useful to examine how shall India achieve the prescribed targets of technology mission on dairy development ? It could be safely observed that there are three basic foundations of dairy development on which the rest of rapid development is envisaged in 2000 A.D. First, a strong infrastructural network built up during almost two decades of the dairy development under Operation Flood since 1970. Apart from the consideration of the benefits of this modern infrastructure, it will be further intensified by the technology mission, which has visualized that almost a "Technology leap" will be necessary; in a variety of areas technology and system will be developed to reduce costs of operation at every stage from the farm to the market" and thus, "technology" will be the main foundation of future development. However, technology could not be developed or used without required investment; adequate invisible funds were made available through operation flood projects. Thus, investment base created by the Operation Flood Project will be another foundation and third and most important foundation of future development will be real instrument of technology transfer in the hands of tiny producers through the most sensitive and reliable rural cooperative institutional network already developed on "Anand pattern" under the operation flood projects. Thus, India's dairy development in 2000 A.D. will derive its strength and support from technology leap, investment funds and institutional network.

Let us draw a sketchy view of the future development likely to be occurred in various areas of dairy sector such as milk production, milk procurement and processing, milk product manufacturing, the milk and product distribution and institutional network.

Though there will be number of dimensions of technological revolution, it is the perception of India's leading dairy stalwart Dr. Kurien that the basic characteristics of Indian production system is unlikely to be changed. This has been explained by the facts that (1) the competition between man and animal for the limited available land will dictate every increasing use of crop residues for cows and buffaloes, while a small limited area may be committed to the production of high yielding fodder, utilization of crop residues such as wheat, rice, straw, stalks and stovers of jowar, maize, and Bajara etc. shall constitute the **back bone of the feeding material** for the dairy animals. (2) Secondly, dairy shall continue to be subsidiary or side occupation for most of the farmers and major source of income for the landless whose number is likely to be increased. (3) Thirdly, the size of the average dairy farming unit also is unlikely to change and the number of dairy animals kept by an average farmer will probably continue to be one or two buffaloes or cows. (4) Fourthly, though, the importance of buffalo as a main source of milk will remain, the cows will find considerable significance as the source of milk production in 2000 A.D.

There will be very remarkable changes in the areas of breeding, feeding and managing of dairy animals at the farm level. However, the considerable burden of promotion of production at the anticipated level would be with "technology of cross-bred animals". In view of a large number of constraints the strategy of cross-bred animals is developed with great care, caution and pragmatism and prolonged national debate on the relevance of such a technology in Indian dairy situation. Today hardly 8% of total bovine animals are cross-bred, though the technology of cross bred animals programme will be wide spread, at the same time. It will be confined to a small portion of the indigenous herd to be replaced by the high yielding animals which are more efficient converters of feed into milk. For example, it is envisaged that under the **National Milch Herd Programme,** it is estimated that by replacing about 1/4th of the indigenous cows by cross breed and 1/5th of the total buffaloes by improved buffaloes the target of the 70 million metric tonnes of milk production could be achieved and to achieve the target of 80 mmt. milk production approximately, 1/3 of indigenous cows and 1/4th of local buffaloes will be required to be replaced by cross-bred cows and upgraded buffaloes. Thus, India's milk production through cross-bred technology is drawn well within the fold of supply of feed and fodder available in 2000 A.D. However, technological advancement made in the area of "Animal breeding through the development of Artificial Insemination Network" during the operation flood project will be well established. India has world's largest artificial insemination (AI) network as it has 30,000 AI centres carrying out 15 inseminations, 60% sperm stations supplying frozen and chilled semen to all AI centres and 200 farms supplying bulls for the semen production stations. However, the number of breedable animals covered under AI is just 8% of the total breedable animals. This implied that a great task of extension of AI will remain an important programme even in 2000 A.D. However, the efforts will also be required to be made to optimise the genetic gain through the new technology of the "MULTIPLE OVULATION AND EMBRYO TRANSFER (MOET)". It is reported that one highly priced cow through this technology of MOET has a potential to produce some 100 calves. Realizing the significance of this advanced technology of animal breeding Government of India has initiated a science and technology project on "cattle herd" improvement for increasing productivity using the embryo-transfer (ET) technology.

The NDDB has also established its main ET laboratory at Sabarmati Ashram Goushala with a donor herd of 162 elite cows and receipt herd of 25 cows. Further, it has a plan to establish 25 state ET centres during the project period. It has been expected though this technology is not available today, may be worked out before the turn of the century. Introduction of GENETIC ENGINEERING by gene transfer for increased milk production is also expected to be one more dimension of technological advance to be recorded by 2000 A.D. (Experience have successfully shown that milk production is also controlled by hormonal levels land the genes which control through hormones.) Exogenous injection of growth hormones is bound to increase milk production by 15% to 30% in bovines.

The technological innovations will be reflected in the area of "Feeding" of the dairy animals also. The simple changes in feeding practices by a small technological device would generate substantial output with cost benefits. One such example is introduction of "Multi nutrient UREA MOLASSES BLOCK LICK" in ruminant feeding practice in India by NDDB. The continuous use of these blocks is bound to increase the intake and digestibility of straws, leading to faster growth of cattle and increase in milk yield with substantial overall reduction in the cost of production by about 30%,; obviously, such innovation in feeding of dairy animals will increase the returns of the milk producers. It is reported that under the operation flood project the co-operatives have set up six UNB manufacturing plants each one of 40 mt./day capacity and many more are under installation.

On the basis of popular practice of using **compounded cattle feed** an attempt will be made to innovate in the area of compounded cattle feed. It is believed that the new generation of cattle feed that would be available by 2000 A.D. will support the milk production at 60% level of feeding in comparison with traditional cattle feed due to its increased capacity of protein conversion, dry matter conversion and decrease in metabolic heat production. Further, the fodder production shortage will be attempted to resolve by attacking on the problems of limited land under fodder production, low productivity of fodder, longer gestation period of fodder production, degradation of grazing land and deforestation of forests. This will be done through suitable technology transfer through well supporting institutional set up. For example, two projects. namely 'SILVI PASTURE and TRE GROWERS' COOPERATIVE SYSTEM' in village common grazing land and other community waste lands have been initiated by NDDB under the operation flood for ameliorating the fodder supply situation and meeting the fuel needs of the country. Under the "SILVI PASTURE PROGRAMME" by March, 1988, 2442 "Kisan-vans" covering 1743 hectares of marginal lands of individual farmers and another 88 GRAMVANS covering 558.8 hectares of communal marginal waste lands were established. This programme has already been initiated in 7 states covering 59 milk sheds.

The Area of **Managing Health of Animals** will also be strengthened by the network of solid health cover in the form of veterinary health services and vaccination services. It is reported that under the operation flood projects as on March, 1988, 394 regular mobile units, 305 emergency mobile units and some 19373 village veterinary first aid centres are already operating. Only Mobile units covered around 3.3 million cases in 1978-88 State governments also provide veterinary services by 15870 veterinary hospitals and 19918 stock maintenance centres. It is envisaged that by 2000 A.D. this minimum animal health facility will be quite commonly available to a largest number of villages in the country. More importantly some of the basic animal diseases like FOOT AND MOUTH DISEASE creating havoc every year in the country leading to the waste of estimated milk production valued Rs. 400 crores per annum will be eradicated through the intensive practice of FMP vaccine which would be universalised and to make this practice more feasible and acceptable, efforts are also being made to develop ORAL VACCINE administered through the "urea molasses lick" which would help eradication programme more rapidly in India by 2000 A.D. It is hoped to develop vaccine to protect exotic and cross bred animals more effectively. A successful experiment to develop "THELEROSIS VACCINE" has created this expectation.

The procurement of milk will become more problematic and challenging due to increased average distance covered to collect milk as a result of intensification of competition and cooperativization. The response will be reflected in the technological innovation; adoption to overcome the problems of distance, transport, climate, seasonality, etc. giving rise to high % of severage and spoilage, twice a day milk procurement system from the village will be exceptionally seen as the innovation of LACTO PEROXIDE SYSTEM will be widely introduced. It may be surprising that some advanced milk co-operative societies in the advanced regions will completely develop a computerised system of procurement. "BULKTANKS" at the society in the village and transport by road milk tankers will be so usual by 2000 A.D. Similarly, the conventional milk measuring instruments will be displaced by the ultra modern electronic system of milk meters measuring and recording milk fat and S.N.F. more accurately with least costs and with great convenience. Electronic system will be used in recording producers' accounts of milk procured and milk payments on the spot.

The rapid installation of the dairy plant processing capacity during the operation flood projects have established a net work of 125 dairy plants in the country with the processing capacity of 128 lakh litres daily. Moreover 'National Milk Grid System' has been developed in such a way that the maximum quantum and largest possible number of towns and cities could be covered by the co-operative dairy system. As a result, though in metropolitan this share is more than 60% and likely to be even 90% by 2000 A.D.Ioday the organised processing sector has its market share of just 12% which will be raised to 50% of milk production in the country. As such seldom any town or city will be left without the benefits of modern dairy system by 2000 A.D. It also maybe noted that today 496 towns including 4 metro cities are already covered throughout India. There are atleast six million consumers who are served with around 100 lakh litres per day. Under this rapid progressive coverage; conventional city dairy stables or traditional dudhia milk vending system and even traditional dairy co-operative societies working around the cities will perhaps start disappearing.

The giant national milk grid system will ensure with best long distance transportation with insulated road and rail tankers. The present rail tankers capacity of 40 lakhs litres will be increased to 108 lakh litres over the next few years. It will also be supported by the giant storage facility as the current storage facility of 11000 metric tonnes for smp and 220000 metric tonnes of conserved fat will be expanded by added storage capacity of 5000 metric tonnes of smp and 2000 metric tonnes of conserved fat even in the near future and thus India's NMGS will be very strong enough to balance milk demand supply system in the whole country by 2000 A.D. Latest technology like GLYCOL CHILLING have already been deployed for long distance transportation without deterioration in the quality of milk. The modern long distance milk transportation through "CONTAINERISED" tankers and even in 'CONDENSED MILK" will result into smooth and stable supply line with least sewerage and spoilage.

The expanding process of coverage of unorganised milk market through cooperativization will result into larger dairy units in terms of their daily capacity of processing, manual system of milk loading at the dairy docks will be replaced by mechanised system with load cells and flow methods to receive milk transported by tankers. Looking to the current crisis of energy; intensified efforts of energy auditing and energy saving devices adopted by the dairies in the cities will be quite usual. The currently initiated solar heating system on experimental level will be completely commercialized and established. The dairy plants will install highly energy saving devices as the energy intensive dairy system will be found most unpopular and much costly by 2000 A.D. It is reported by the experts that energy saving consideration may lead to innovation in regeneration of heat in pasteurization. The dairy technology in India will be capable enough to reduce the fluent treatment costs, waste to milk due to drippings and careless handling. The use of water and other utilities will be minimised while the recycling of effluents will be adopted to maximise the benefits. The use of water and other utilities will be minimised while the recycling of effluents will be adopted to maximise the benefits. The computer system will be common not only for the technical operations and quality control at the plant level but also extensively used as the device for the sound information system at each dairy plant with its package to national computerised system of information network under NMG which will help regularisation and stabilization of the demand and supply of milk products all over India.

The Milk distribution system by 2000 A.D. will be based on modern technologies adopted in advanced dairy countries. The concept of Ultra High Temperature (UHT) milk has revolutionised the liquid milk market even today. UHT milk has double advantage of availability of milk throughout the day and enabling the feeder dairies to enter the urban market without creating additional capacities in major market centres itself. (Ascetically packed UHT milk is expected to account for about by the co-operative sector itself.) (In fact, revolution in the milk distribution could be seen in the trends of substitutions of the forms of distribution of milk since 1985. The share of the sachets, cans, bottles, bulk vending machines in 1980 was 50.0, 18.5, 17.5% and 0.72 which has been expected to be increased by 1990 at 50.0%, 5.3%, 5.0%, 30.0% and 9.7 share of sachets, cans, bottles, bulk vending and aseptic respectively.) Obviously, the further trend of these milk distribution forms would mean that vending machines, tetra pack or pouches will be so common that the glass bottles will be relegated to the past.

The milk utilization pattern will be still dominated by the liquid milk as it was in 1951-52 where the share of liquid milk was 39.3% and 'Ghee' had also almost equal share of 40% while rest of the 20% milk was used in the form of curd (8.8%), butter (6.0%) and Khoya (4.4%). This traditional pattern of milk utilisation has changed due to modernisation and changes in the demand. (Though, liquid milk has continued to dominate, it constituted 46% share in 1955, significance of ghee has declined while the new product lines such as milk powder and cheese have started assuming significant place in the milk utilization pattern.) This trend of changing milk utilisation in the fold of modern milk products will be continued to reflect the increased demand and changed demand pattern due to the factors such as increased income of urban and rural households, increase in the life expectancy, decrease in the size of the family, growing automations in home kitchen wider use of refrigerators, growing tendency for the out-door eating, an emphasis on quality, increased consciousness on the health aspects of food consumption such as slimness, fitness, energy, low cholesterol, low fat etc. In fact all these factors will shape the future pattern of milk product in 2000 A.D. Some products like cheese and milk powder will have much brighter future. For example, it is expected that with the growth of fast food joints, the cheese seems to have maximum growth potential during the 21st century. Increasing quantum as well as range of cheese varieties might promote the growth of products. (Though the increased rate of production of milk will certainly help stable supply; in view of the progressive production of even gifted imports of butter and butter oil, significance of indigenous skim milk powder to balance the regional and seasonal fluctuations in milk production and to recombine milk for the production of toned milk etc. will increase.) In fact, looking to the current rate of production of milk and buffer stock we may become net exporter of the milk power by 2000 A.D. However, the most interesting will be the growing demand and supply of indigenous milk products. A very successful technological innovation of the production of "SHRIKHAND" has started establishing a new trend for indigenous milk products covering all indigenous regionally dominated products such as gulab jamun, Rasgulla, Pedda, Barfi. These products will find a national market and will find a place in product mix of the major dairies. Obviously, this will help profitable disposal of the surplus ultimately leading to better returns in dairy enterprises.

Finally, the institutional development process initiated under the operation flood project since 1970 will be replicated in almost all areas where there exist dairying potential. It is estimated by experts that by 2000 A.D., we will be in all probability of around 80,000 village milk co-operatives working in 200 milk sheds covering 275 districts of India. The recent studies have also shown that the Anand pattern dairy co-operatives are fast replicating in non-Gujarat regions who have greater growth rates per annum. There are atleast six states which greater growth rates per annum. There are atleast six states which are very leading states such as Gujarat, M.P., T.N., A.P. and Rajasthan. Thus, Anand pattern will be all India pattern by 2000 A.D. Similarly, all states will be having active state federations on the line of Gujarat state with milk marketing federations. Moreover, recently established all India apex body of National Federation of Dairy Cooperative will start playing pivotal role in co-ordination, regulating and thus strengthening the national milk grid system in a real sense. Similarly, it is very likely that present co-operative dairy enterprises will not only develop commercial outlook like corporate enterprises but also will assume giant form and size and thus, if unless restrictive practices are adopted, they will take the shape of "co-operative giants" of USA and Canada. The current status and size of "Amul Dairy" and other fast co-operative dairies in Gujarat provides a base to anticipate such a possibility by 2000 A.D. Amul dairy has recorded a turnover of Rs. 200 crores and thus established as the largest co-operative unit of the whole Asia. Thus, Indian dairy sector will assume a very significant place in the agricultural sector, agro-industries and rural industrial enterprises in India.

III

Dynamic dairy development in 2000 A.D. will be characterised by the changes always well woven with the challenges. A pragmatic response to the challenge will only help us to make up opportunities of growth for the sustainable development. The process of dairy development for 2000 A.D. is likely to raise the following major problems.

As we have seen the secret of the quantum jump in the modernization almost destroying. "Dualism' of technology could be possible due to well designed and well implemented projects of the operation flood. However, one of the reasons of the smooth sailing of the process of the development was continued support in the form of gifted dairy aids and liberal loans of the World Bank atleast after the II phase of the operation flood project. As we shall see presently, such a continuous flow of the international support neither could be continued for ever nor could be desirable. It is quite appreciable to note that almost 1/3rd of the outlays of the Operation Flood phase-III was financed through internal resources raised leading to progressive reduction of international dairy commodity aid. However, we may not have a problem of investiable funds perhaps in the near future; for example, Working Group on dairying and animal husbandry for VIII plan has documented that it is estimated that the spill over funds of operation flood III available for the deployment during the VIII plan (1990-95) would be a sum Rs. 6430 million. Apart from the needs of resources to develop the further phase of the project, one could see resources needed for the development. For example, the Working Group itself has noted that as in the earlier plans, dairy development activities would continue to be undertaken in the state sector with special focus on the aspects of **technology mission** (as we had also shown earlier). It is estimated that continuation of such activity in the state sector would require Rs. 3000 million. Thus, the decade of nineties is full of development environment but it has to be supported by the investible resources which are not easily possible to manage as compared to decades of seventies and eighties.

The main reason of slow but certain withdrawal of dairy aid could be explained in terms of adverse international environment also. International dairy aid of EEC was mainly possible due to mounting surplus of milk and products which had a dumping effect on the price level as a result, its quick disposal was an economic necessity. This scenario of international dairy industry has been substantially reversed. Today as compared to 1970 and even 1985, there are not much surpluses and as a result the price of skimmed milk powder which was 200 $ per tonne in 1985 had increased to 600 $ in 1989. The price of milk fat increased over 60% over the same period and surplus stock of powder and butter oil have almost disappeared. It is also reported, that as a result of this changed scenario, the availability of milk powder for food aid has dropped from around 400, 000 tonnes in the early eighties to 100, 000 tonnes in 1989. This dramatic change in price and supply has serious implications for developing countries like us where dairy industry has already been developed but equally developed dairy farm sector to a large extent is yethas not been possiple.

However, the current strategy of operation flood to concentrate on indigenous production of milk powder facilitated by increased rate of milk production and increased capacity of manufacturing milk powder will be emphasised. It is reported that milk powder production had gone up from the level of 22000 tonnes in 1976-77 to 100,000 tonnes in 1987. This provides great opportunity to develop BUFFER STOCK to offset basic constraints of Indian dairy sector such as a very erratic seasonality, regionality and instability in production of milk faced by the dairy industry. The seasonality could be seen in the fact that 2/3 of the total milk is produced in this country in a span of four months and 1/3 in the balance of eight months while demand is relatively stable. An extent of instability could be explained by an example. In 1986, an average of 8 lakh litres of milk was rejected by the dairies because limitations of the storage, processing and marketing facilities. In the two decades production of milk increased by 20% a year but such a rate of substantial increase was halted by the prolonged drought in the milk surplus states, regional imbalances could be seen from the fact that 90% of the production came from 12 states out of 22 states of India. To put the picture of regional imbalance more sharply, we may say that the western regions contribute hardly 3% of national milk production. It is also important to note that the conserved commodities are required not just for balancing demand and supply of milk as seen earlier. Skimmed milk powder (SMP) and butter oil are needed for the purpose of STANDARDIZATION of milk which will increase due to the reasons of high price, health consciousness and habit formation of the consumers in 2000 A.D. This calls for the building up of the bufferstock. Recent trend of the milk powder is quite satisfactory. In fact, it has been reported in the press that we could provide for some 'surplus' to export to developing countries and import of edible oil in exchange of these exports.

The increasing surplus milk products would lead us to be an exporter. International trade of milk products like milk powder and some other Western products like cheese is not impossible. This could provide us a great source of required foreign exchange to import essential technology. Though, the international trade with developed countries would be very much challenging on account of high cost of production, low quality of product and also lesser demand as developed countries will find the milk production quite stable on account of new technology. It is reported by the planners in the countries that it is expected during early 21st century, that there would be stable milk production in the advanced dairy countries and expansion in the developing countries but there will be higher international market price due to the sharply reduced surplus; a decline in number of dairy farmers and the number of persons employed as a result of technological progress in the advanced countries. Whereas neverse changes are envisaged in the developing countries; the world over there would be greater priority of health, environment and alternative methods of milk production, increased international inter dependence through trade and growth in organizing marketing and uniformity in dairy polices. All these factors will provide Indian dairy Industry, a "lead role" as Indian dairy industry will certainly be much advanced than many other countries of Asia. Recent request of the various countries like Nigeria, Ghana and Madagascar through World Bank for the development of dairy industry on the line of Indian operation flood project, provide a great opportunity to develop "South-south cooperation" in international trade of milk products. The South Asia and some African countries and oil exporting countries could provide a great market to India for the export promotion, provided quality is maintained at international standard and at competitive costs. Thus, export of the 'surplus' milk products of India in 2000 A.D. is a great possibility and all efforts must be made to explore these opportunities.

The great challenge Indian dairy industry will have to face is on the side of cost of production of milk and milk products. The increased supply and surplus and output at increased cost will create the problem of disposal even in Indian market. It is estimated that price trends of the consumer price of dairy products will be atleast by 2000 A.D. Baxi, of Gujarat Milk Marketing Federation has estimated the average approximate price of milk and products in India in 2000 A.D. as follows:

Year	Milk-Rs/litre	Ghee Rs./Kg.%	Butter Rs/Kg.
1990	7.00	70.72	63.65
1995	8.50	80.85	70.75
2000 A.D.	10.00	110.12	100.11

(It is notable that average price rise of each product is more than 6% rate of inflation in 2000 A.D.) There are evidences of research to show that high cost of production have led to decline has the expenditure on milk and products.

The only option open to cut cost of production is adoption of 'technology' which will ensure reduction of costs with improvement in productivity. Comparative economics of cross bred animals and local animals shows cross bred could produce litres milk per day with 0.8 Rs. per litre cost while local fed animal could produce just 1.2 litres of milk at 1.0 Rs. per litre cost, thus reduction of cost and productivity both are achievable simultaneously through cross-bred technology. However, the experience of the operation flood project shows that the adoption of cross bred animals is still rather very slow. The studies have recorded improvement in productivity also but this productivity rise is much less attributed to the cross-bred animals. A survey of IIPO (1984) reveals that cross-bred accounts for only 4.3% of the average number of milch animals per household, while their number has been rising, their contribution to increase milk production so far can be regarded as only marginal. The study attributes the productivity rise to increase number of producers and improved feeding practices and not to the cross-breds. In fact, the cost of co-production has not declined even in those areas where production performance is notable. Shows that much is to be done in this direction.

The established case of technology of cross breeding itself would pose some other problems to resolve. First increasing difficulties of participation of marginal producers and landless producers on both the counts, i.e. initial investment, local buffalo priced Rs. 3000, upgraded buffalos Rs.13,000, local cow Rs. 2000 but cross-bred cow Rs. 10,000, plus there are problems of adequate supply of cross-bred animals. Obviously, at premium price better off non-poor producers would take over the weaker producers and equity dimension at the dairying will be tilted in the favour of non-poors. A series of measures will be required to undertake during this decade to see that this class of producers also could take advantage of new technology and not bye passed . Otherwise like green revolution growth will be achieved without social justice.

(The cross bred cow yielding 7 to 8 times more than local ones, are better converter of feed but also need to have more feed per cross bred. For example, against 10 kg. per average local fed animals, it needs 20 kg green fodder per day, and support of concentrates atleast 2 kg. per animal per day against just 0.15 kg per animal.) This increased demand for the fodder has to be related to our current fodder availability also. The picture for green fodder is depressing as can be seen from the following :

Year	Permanent other land pastures and grazing	Cultivable waste land
1960-61	14.0	19.2
1965-66	14.1	17.0
1970-71	13.0	17.5
1975-76	12.0	17.4
1980-81	12.0	16.7
1981-82	12.1	16.4
1982-83	12.0	16.3
1983-84	12.0	15.5

Source : Statistical Abstracts 1986

The standard 4.0% land under fodder production is found unchanged. However, it is likely that some increase might have occurred in Punjab and Haryana. However, one may anticipate that increased demand for fodder may induce the larger farmers to devote some land for fodder also. However, a systematic drive of social forestry programmes to enhance availability of fuel wood as well as fodder particularly under NREP and RLEP must get rid of the problem of access to community land and must learn to develop participatory of development. If IRDP is linked up with co-operative structure of PROJECTS both could contribute in major way.

Finally, the challenge of development also lies in the management of 'Change'. The managerial challenges will arise at both the levels, at dairy farm level as well as dairy industries level as soon as the DUALISM between these two sectors will start disappearing through technology leaps as envisaged by the development in 2000 A.D. The management of cross breed cattle would be formidable challenge to farmers. The provisions of training and extension will have to be made to train producers in breeding, feeding and managing these precious animals. At the level of fast modernizing dairy sector through co-operative process, the need for peculiar class of "Rural Managers" will arise. This class of rural managers would be well trained professional conversant with modern management like any other common professional but they will have to distinguish them with others in the one sense that this class will be totally committed to "rural cause" and "rural community". Thanks to the dynamism and vision of India's greatest dairy man Kurien who has already realised this very essential condition of development and therefore, established the Institute of Rural Management (IRMA) but the moot question is that shall we be able to cope up with available meager supply of such a IRMA's when India will be needing much larger number of such a class managers. Unfortunately, nowhere we are finding ourselves to prepare to meet this challenge of managerial manpower needs. The accelerated efforts are required to replicate the IRMA mode to satisfy the manpower needs of modern dairy sector in 2000 A.D. Thus, co-operative development prospects are quite brilliant in 2000 A.D. However, an extent of the potential will actually be converted into progress and will very largely depend upon our capacity to combat the critical issues arising out of the development process which has already began since 1970's. We can pin our hopes on our past experience of achievement and expect that we shall be able to convert every critical issue into clear cut opportunity of the development.

References

1. V.Kurien, Dairying in India, 2000 A.D. (Keynote Address to XIII Dairy Conference, 1988).
2. M.P.G. Kurup, Technological Innovations in Milk Production (International Dairy Seminar, Paper presented in Dairying as an Instrument for the progress. IDF (NDDB 1988).
3. P.C. Bardhan, Operation Flood - Its evolution, the strategy followed and progress achieved (International Dairy Seminar paper).
4. I.Z. Bhatty, The Role of Dairying in the Agricultural Economy of India, (International Dairy Seminar paper).
5. J.J. Baxi, Dynamics of Marketing of Milk and Milk Products in 2001 A.D. (IDC Conference of 1988).
6. Dairying for 21st Century-XII Dairy Industry Conference, 1988. Recommendations.
7. A.K. Ray Chaudhari, Balancing Milk Supply and Demand Proceeding of the Seminar IDM, 1988.
8. Dr. K.K. Iya, and Dr. B.N. Mathur, Dairy processing Industry of 21st Century, Scope and Visions, IDM - 1988.
9. Nishit Kumar, Communication strategies and Needs of the dairy industry in 2000 A.D., DIM - 1988.
10. R.P. Aneja, Meeting the demand for milk and milk products in the 1990's - Seminar paper, 1989.
11. R.B. Singh, A. glance at Operation Flood - Phase - I, II & III IDN, 1989.
12. V. Mukund Das, Future Targets in Milk and Milk products marketing - Needs for strategic directions, Indian Dairyman Vol. L. XIII Bov. 4, April, 1990.

13. V. Mukund Das, The Emerging marketing Environment for milk and milk products, IDN, 1986.
14. C. Srinivas Sastry, Scope for South - South.Cooperation in Dairy Development (International Dairy Seminar, 1989).
15. Policy formulation for dairy development in Asia, An International Workshop report, 1989.
16. R.S. Gill, Proceedings of IDA Seminar on Balancing of milk supply and demand.
17. Peiter, M. Hoogendijik, Technological Innovation in Milk Processing - Summary of the lecture presented at International Seminar on Dairying as an Instrument for Progress (international Seminar Paper - 1989).
18. D. Tikku, Innovation in Milk Distribution, (International Seminar, 1989).
19. Zecharich, Mathews. Innovative Technology for Animal health Care XXII, Dairy Industry Conference Paper.
20. J. J. Baxi, Role of Innovative Distribution in Developing markets.
21. S.G. Zanwas, Embryo Transfer in Cattle.
22. J. J. Baxi, Emerging Trends in marketing new products launch - A recent experience.
23. J. J. Baxi, Marketing Strategy in operation flood concept of the national milk grid and the mother dairy.
24. S.C. Jain, Planning for manpower requirement of Dairy Industry in India.
25. A. K. Chaudhari : Research needs for developing dairy industry.
26. Dilip Shah, Cooperative Dairying - Part- I and II, Economic Times, 16-17 January,1989.
27. Dilip Shah, Anand Pattern Dairy Development, Financial Express 2-3, Feb., 1990.
28. N.D.D.B. Operation Flood - Phase III - A proposal Sept. 1985.
29. Proceedings of International Seminar on 'Dairying as an Instrument for progress - The Indian Experience NDDB, 1990.
30. Report of the working group on Animal Husbandry (Including Poultry) and Dairying for the formulation of the Eighth Plan (unpublished).
31. Ministry of Agriculture, Government of India, Technology Mission on dairy development, March, 1988.
32. Dairy India - 1987, New Delhi, 1986.

□□□

10

Liberalized Dairy Industry Policy Dairy Industry Scenario in Post Liberalization Period and Strategic Policy Implications for Next Millennium

Introudction :

This paper is an effort of continuous research to observe the changes in Dairy Industry scenario due to major changes in the dairy industry policy of India since, 1991. In an our earlier study (Dilip Shah, 1992) on the policy of delicensing of dairy product manufacturing we observed that expected benefits of growth, investment, investment, technology and export on account of entry of private sector are unlikely to be achieved and thus, liberalized dairy policy would be more harmful than otherwise that study was undertaken prior to the advent of Milk and Milk Product Order (MMPO). To examine the micro-level impact of the liberalized dairy policy, we had conducted two micro-level studies in Post MMPO period with the support of NDDB and IRMA. First study had assessed the impact of entry of Private Dairy operators in Mehshana Milkshed of Gujarat (Dilip Shah, April, 1996). Another micro-level analysis related to Baroda Dairy Co-operative of private dairy operators in Vadodara district Milk Market and City Milk Market (Dilip Shah, 1996).

Thus, More complete picture of the dairy industry liberalization in the dairy sector was obtained as both studies concentrated on Rural-urban experience of dairies which were product market oriented as well as Liquid Milk market oriented in their dairy operations. The present paper in attempt to analyse the more recent scenario of liberalized dairy industry and strategic policy implication for the dairy development in next millennium.

The present paper is designed to deal with all relevant aspects of the theme in the following sections. To provide a back ground the first section briefly high lights the evolution of dairy industry policy in India. Second section, based on available latest data attempts to depict the Post liberalization scenario of Dairy Industry. For justified coverage, we viewed the Dairy Industry as a system comprising three inter-related subsystems namely Milk Production (Katarsing, 1997). Same section documents the evidencees based on our studies as well as other studies to records the consequences of the liberalized dairy policy. The 3rd and final section in the context of our analysis in the second section explains the strategic policy implications for the next millennium for the rapid, balanced and desirable development of dairy industry. The paper ends with an appropriate concluding observations.

I. EVOLUTION OF DAIRY INDUSTRY POLICY

Review of the post independent India's dairy Industry Policy shows three distinct phases:

1. Government sector oriented Dairy Industry Policy (1050-70).
2. Co-operative sector oriented Dairy Industry Policy (1971-90).
3. liberalized Dairy Industry Policy with opening up of Private sector entry in organized dairy processing, product manufacturing sector (1991).

1. Government Sector Oriented dairy Industry Policy:-

Though Government had recognized the need for milk production promotion in view of the increasing population and income. It was highly "Urbanilised" subsidized policy which led strategic failure in the sense that during the period of 1951-1970. It led only marginal rise in Milk Production leaving for major imports to feed the metro-plants installed. During 1951-69,70 average rate of growth of Milk Production was less than 1% markedly inadequate to meet the increasing population at the rate of 2.6% average rate of growth. The stagnant productivity of milk led reduction in per capita availabilities from 132 gm for 1951 to 107 gm in 1969-70. Consequently, the use of imported Milk material increased substantially i.e. Skill Milk Powder (SMP) from 6700 Mt to 27400 MT (+%), butter oil (BO) 500 MT to MT (+ %) (Dairy India - 1997-98), however, by 1970, the Central Government had accepted the Anand pattern dairy Co-operative structure as the best farmers' organization for the development of dairying through small farmers, marginal farmers and Inandless labors us recommended by national Commission on Agriculture, 1976. This acceptance of strategic policy change implied a remarkable reversal of policy focus from urban to rural orientation providing priority to production along with simultaneously promoting of processing and marketing facilities through integrated co-oprative dairy structure as per Anand co-operative model.

II. OPERATION FLOOD PROJECT BASED CO-OPERATIVE DAIRY INDUSTRY POLICY:

Operation flood project designed to replicate Anand pattern dairy co-operative through National Dairy Development Board headed by V. Kurien has a long vivid history of its origin and evolution (Dilip Shah, 1982), however, we may briefly state that this world's largest dairy development project through its three phase Ist, IInd and the IIIrd Phase of the operation flood has been completed in 1996. Phase IVth of the pogramme has been implemented on a modest scale by NDDB. Out of its own resources, this period of 1970-1096, 25 years of dairy development through the operation flood has always remained controversy of points and counter points (Dilip Shah, 1996), however, two tables in the Appendix are self-explanatory document to show how operation flood had worked for changing the basic structure, status and strength of the dairy industry of India. It must be noted that even the World Bank has appreciated the substantial contribution of the projects. In its latest and final evaluating report of the world bank has also now endorsed the achievements of the operation flood projects. It has accepted the O.F. Projects led revolution from green to white, though it took time to endorse that the there has been real rise in Milk production and yet its study do not give full credit to operation flood on the ground that "the actual volume of the milk handled by operaion flood remains a small fraction of the increased supply, i.e. 6.3% of total milk production or about 22% marketed surplus. It is thus not possible to attribute increased production simply to the dairy processing, marketing infrastructure and technical support provided by Operation flood directly Important though these contribution has been rather, one has to look to the changed dairy policy environment which accompanist the decision supporting Operation flood. However, World Bank evaluation is ar more firm in the two areas of achievements of achievements of operation flood projects one is self-reliance by "Indigenous dairy equipment manufacturing"; which has contributed by 93% i.e. hardly 7% dairy equipment are imported. Though this is regarded by the World Bank as "A Byproduct of operation flood" and thus, reduced its significance. To indicate the project based responsibility, it has rightly quoted the statement of Dr. Kurien "Operation flood is not an all purpose poverty removal programme" and then, it has documented the clear benefits for the poor and women.

For those landless people who own or have been able to purchase a milk animal is cow or buffalo. It has been a boon of course, those who are both cowless and landless have only been slightly affected by some of the social programme cooperated by the villages Dairy, Dairy cooperative societies. Out of their porfits and by the increased prosperity of the village economics. Operation flood projects in collaboration with NGO's has established 6000 women's dairy cooperatives societies and generated higher incomes from some women. indirectly. It has expanded the number of children attending school because of the high income elasticity of demand for education in Indian villages. This project has produced diverse results on the ground. It has estimated that it could create additional 1,75, 000 laboring jobs, predominantly for very poor and it has led additional 3,50,000 extra children in school. Thus, the World Bank has appreciated very significant contribution of the Operation Flood Projects in developing dairy sector along with more significant socio-economic benefits with efficient management of rate of return on investment. However, those who have benefit of consulting other independent studies would agree that perhaps the World Bank has "Underplayed" the achievements of the operation flood projects.

III. LIBERALIZED DAIRY INDUSTRY POLICY

'Liberalization of the dairy industry was not because Operation flood failed' (Kurien, 1997). This is relevant to note as liberalization is prescribed by the World Bank of India. In fact, a recent World Bank study find the World Bank during the last decade has been no friend of co-operatives or the N. D. D. B and operation flood - suggests that 'India's milk production is as much as 40 million tons per year higher than it would have been without operation flood. But Governments seem to take success for granted , and then remove the very basis for that success (Kurien, 1997).

A fundamental dairy policy departure was made against the national policy of dairy development through, Anand pattern dairy co-operative structure. As a part of Economic 'liberalization' dairy industry is also liberalized in July, 1991. We have discussed its logic and reality in our earlier paper, however, this is great policy departure by way of exempting dairy processing industry from industrial development and regulation Act of 1951 which formally prevented the entry of private dairy operators and companies in the industry.

Another policy change in the favor of multi-nationals was that they were allowed to raise their equity holding up to 51% gave an extra ordinary strength to all present companies of MNC in India to raise the industry is also liberalized in July, 1991. We have discussed its logic and reality in our earlier paper, however, this is great policy departure by way of exempting dairy processing industry from industrial development and regulation Act of 1951 which formally prevented the entry of private dairy operators and companies in the industry.

Another policy change in the favor of multi-nationals was that they were allowed to raise their equity holding up to 51% gave an extra ordinary strength to all present companies of MNC in India to raise their equity holding, Glaxo and Nestle who had not more than 40% increased their quality holding to 51% and then, strengthen their base and interest. The very first consequence of the industry were found in terms of numerous plants came into existence without adequate desirable hygienic standards, most of them meant for product manufacturing only. Moreover, co-operatives strongly opposed the policy on the ground that private units are taking advantage of ready crop in terms of milk production and procurement for which co-operatives had done huge monetary and manpower investment. In fact, co-operatives felt truly threatened to the fact that out of 689 applications, 95% were for product manufacturing. Some state like Punjab even attracted Private Investment by liberal incentives as a result, capacity exceeded the potential production and procurement. In fact, this over concentration of manufacturing products, excess capacity and the malpractices led the Government of India to notify The Milk and Milk Product order on 9th June, 1992. As per the provisions of this order any person/ dairy plant handling more than 10,000 liter per day of milk as 500 MT of solid per annum, needs to be registered with the registering authority by the Central Government. The provisions also included obligatory liquid milk sale and also the development of their own milkshed etc. Thus, MMP was empowered to control an entry and operation in liberalised dairy sector.

World Bank in its livestock sector review of India had expressed its perception indicating that dairy sector of India had received continued protection leading to inefficiency in processing industry and industries development and regulation Act of 1951 was limiting the competition from the non-cooperative private sector. In the context of liberalization, the licensing requirement was abandoned in 1991; only to be reintroduced in 1992 under the MMO, in response to political pressure. Thus, backdoor entry of regulation was done.

On 27th August, 1993, the MMO was amended authorizing the State of Government to register the units handling less than 75,000 litters per day or it is quite notable the different views are expressed by co-operatives and liberalization advocate World Bank. On the one hand, the co-operative dairy sector regarded modified MMPO as "TOOTHLESS" as the diluted MMPO registration requirement is abolished upto 75,000 litters milk or 3750 MT milk powder and also renewal term period also increased from 3 to 5 years and advisory body was also reconstituted to regulate the strength of co-operatives. It was feared by NDDB that the capacity creation has already exceeded the quantum of Marketable surplus.

On the other side, World Bank had expressed its dissatisfaction against amended MMPO of 1993, as several key regulations were retained which include requirement that new processors must develop the milk sheds. It was feared by NDDB that the capacity creation has already exceeded the quantum of Marketable surplus.

On the other side, World Bank had expressed its dissatisfaction against amended MMPO of 1993, as several key regulations were retained which include requirement that new processors must develop their own milkshed or milk collection area and cannot encroach on cooperative Milkshed and processing of milk in to higher value products could be banned during the lean period.

The World Bank had argued against the MMPO on the ground such regulation would restrict the entry and thus competition and growth. It also discriminate against achieving economies of scale and technology modernization to improve quality which would help the industry to become competitive. It was also required that MMP would lead to : a) higher market price of products b) defective MMPO's will restrict production and also will lead to high processing cost and also inefficiency of processing i.e. processing cost in India was already reported to be 20% to 50% higher than international average standards.

DAIRY INDUSTRY SCENARIO IN THE POST DAIRY LIBERALIZATION PERIOD

We have attempted to sketch the post dairy liberalization scenario at the national level in the selected states and at micro level rural urban levels in Gujarat State which is still leading with almost 30% procurement of total production against just 6.67% in all India indicating almost 5 times greater progress in the co-operative dairy industry as compared to all India.

All India level MMPO status as on 1995 was as follow :

	Co-operatives	Others (Non-co-op.)	Total
Applications (No)	256	442	698
Existing Capacity (LLPD)	253	169	404
Incremental capacity (LLPD)	73	465	539
Total capacity (LLPD)	308	634	942
Rural Milk Marketable surplus	-	-	456

Source : Informal

Regional distribution of the applications recorded highest application in UP (174), followed by Punjab (57) and Haryana (71), Rajasthan (41). thus out of 698 43% were belonging to these states.

The table shows clearly that over the five years, capacities have been registered exceeding the available marketable surplus for the processing and this excess capacities will make our processing efficiency more inefficient as our industry has processing inefficiency compared by international standards.

One study shows the MMPO operation in Gujarat with reference to All India in more recent period (1996 and 1997-98) (Cap. 000/LPD).

	Cooperative		Private		Other		Total	
	No.	Capacity	No.	Capacity	No.	Capacity	No.	Capacity
(A) Central Authority								
Guj.	14	6170	1	150	6	640	21	6960
All India	124	21580	73	12664	30	5390	227	39634
(B) State Authority								
Guj.	03	120	00	0	1	20	4	140
All India	56	1832	140	6462	5	185	201	8479
(B) Total								
Guj.	17	6290	1	150	7	660	25	7100
All India	180	23412	213	19126	35	5575	428	48113

Source : P.N. Thakar, A. J. Pandya, Dairy Industry of Gujarat - A profile IDA - 1997.

It may be noted that handing capacity of Gujarat is already 66.70 LPD while milk procured is 31.55 indicating gap between installed capacity and utilization under this situation. We could clearly justify the conscious behavior of state authority granting licenses to the new plants in Gujarat.

The latest Annual Report of the department of animal husbandry and dairying ministry of Agriculture for 1997-98 shows that the Central registration authority has granted registration for 241 units with combined capacity of 437.20 lakh litres per day, upto December, 1998-99; since the inception of the order, 225 applications have been rejected on various grounds, mainly for want of surplus milk in the milkshed area and from hygiene and sanitary conditions in the plant premises (Department of Animal husbandry and Dairying, 1997-98).

One of the major consequence of the liberalizaed dairy industry is large scale production of synthetic milk to meet increase demand by private dairying, reports from U.P., Punjab, Haryana and Rajasthan where highest numbers of applications were granted for Milk Processing and Production Manufacturing. "Milk Mafia" has almost played a havoc (NDDB, New Bulletin, June, 1999). Synthetic milk has ingredients like urea, detergents, caustic soda, vegetable oil, animal fat, sodium sulphate and even formalin and white paint. These ingredients in milk may cause damage to vital body organs such as liver, kidney, and heart (Hindu business line, Thursday, June 1999), however, now it is reported that anew method to test synthetic milk has been developed to detect the presence of such ingredients by scientists of Haryana Agricultural University.

It is very interesting that at least at all India level, "Concerns - including those within the Dairy Board - that a rush of private sector and multi-national investment would swamped the co-operatives dairy sector proved prematured (NDDB, Annual Report, 1996-97). NDDB Reports that the co-operatives are doing well as it is evident from the significant improvements across a wide sprectum of activities i.e. Milk procurement grew by 12% reaching about 4.5 million tones for the year with about Rs. 40,000 million going into payment to producer members, procurement increased by more than 20% among co-operatives in Rajasthan, Bihar and Gujarat competition is being met by improving performance. Increasing members of unions have achieved and maintained financial surpluses. Marketing effects have intensified both our major cities and elsewhere. Services to farmers have expanded and improved with more feed fodder, seed, artificial insemination and better veterinary covers all supporting efforts to increase productivity and incomes (NDDB, Annual Report, 1996-97).

It is very interesting to note that "The 720 crores Hindustan Level Ltd. (HLL) has formally put its Rs. 360 crores dairy product and animal feed business on the block. The consumer goods giant had, last year said it would be getting out of those two business segments. It is learnt that HLL's Anik faced tough competition from the NDDB's Amul (Economic Times, 1995 HLL puts animal and milk food business on sale, 12, Jan. 1999). In this context, it is very interesting to note Dr. Kurien's explanation on this 'unpredictable result after liberalization Dr. Kurien says as follows :-

"The second result was not so predictable. It was my prediction but not many others shared my view, most of those firms that have been set-up have been collapsed on all on the verge of collapse. The reason - without getting into great detail of detail about the number of the dairy industry - is that they failed to recognize and account for some of the very important competitive advantages of co-operatives. When it comes to the procurement and processing of milk... Now we have faced and met much of the competition in dairying Amul is bigger in India than 'NESTLE' Bigger than Cadbury and even Glaxo .. (Kurien V., 1997). However, despite this initial success in defeating the completion from India based MNCs); it is still a matter of concern if giant multinational like KRAFT enters Indian Dairy Market. The size and capacity of the MNC could be seen in the fact, that most successful for Amul annual turn over of Rs. 1600 crores (US $ 450 million) is less than 30 days profits of the Philip Morris corporation parent body of 'KRAFT'. So if KRAFT decides to enter India with interest free capital for the Indian subsidiaries, which some multi-nationals are now doing - then we would have a real fight on our hands (Kurien, Ibid).

During post liberalized dairy industry period environment is more clearly opened up in the area of International trade.

There are three sets of interesting features. Peculiar feature in the area of imports, exports ad non-tariff barries extended by leading countries to discourage developing countries like India to take advantage of the International market.

It is very notable that India has committed to zero percent tariff for imports of milk powder, both SMP and WMP. There are hardly any other major country in the world that have bound imports of SMP and WMP at zero import duty except Singapore and Japan has bound SMP imports at zero imports only for school lunch (Gulati, Vandana Raheja 1997). In fact, this zero bound rate for as many as 18 agricultural items was negotiated way ago with the Genera Protocol of 1947. A period when India was used to be perennially dependent on food and programmes and on duty gifted commodition made no sense. The strategic thing is that our policy markets have not negotiated the bound rates since then keeping in view the growing food self sufficiency, even during the 1994. GATT negotiation nobody bothered to negotiate. As a result we are the only country that has agreed to a zero duty on milk powder for the base year 1995 as well as on the final year (2004).

This commitment of India with WTO can not be negotiated till WTO ministerial conference to be need early December, 1999; till then it would not be possible to levy any tariff on imported milk powder. Recent depressed price of SMP in global market as well Indian industry with 12% and 41% respectively between Jan. 1996 and March, 1999 has already depressed the Indian SMP price as the local wholesale price was around Rs. 67 per kg. is almost same with Rs. 3 to 4 variation. Any more dumping of import is sure to adversely affect the interest of producers about which even World Bank has no doubt. "According to World Bank", while full trade in dairy products would benefit dairy consumers, milk producers in the country would not gain and thus, NDDB's position of protection against Dairy imports has been endorsed (Business Times, The Times of India, Feb. 1999). This is a likely cost of dairy liberalization to be paid by Indian Dairy Industry unless at least immediately quantitative restriction of 1995 are restored (Harish Damodaran, "Dairy Industry feels the heat of global trade pressure." The Hindu Business line, April, 1999).

Recent liberalization policy is extended to export promotion by special incentives and Dairy Industry has also got its due share (Dilip Shah, 1997, P.K. Singh, 1995). Apart from decanalization of exports and special promotative role played by APEDA; it removed by "Pre-shipment Inspection etc. Export incentives are redesigned but export licensing and export ceilings maintained with APEDA. Available data from APEDA and other sources show the following main performance features of Dairy products exports in terms of its nature trend, value, quantum, pattern and area of markets etc.

1. Export of the dairy product has been conditioned by our net marketed surplus which is uncertain and erratic.
2. Though the quantum of export in terms of total production was not significant yet, it has shown significant rise in its post liberalized period though the growth rate was erratic.
3. Major exportable products with substantial quantum are two namely butter and cheese and milk cream and sweet cream.
4. Our main export markets are Russia, U.A.R., Bahrain and Bangladesh.

Thus, export scenario of India clearly shows a modest initiation of exports of selected seasonal products concentrated in some developing countries with erratic trend.

In the context of the Europe and USA and other country, India enjoys some provision price competitiveness and the implementation of GATT/WTO of removal subsidies should provide much better opportunity for Indian dairy products exports, however, in reality it may not be so due to non-tariff barrier which are already created to deprive the opportunity of export for country like India.

We may record the latest report on non-tariff barriers from what Dr. Kurien has stated.

"These countries are now busy figuring out how to set-up new tariff barriers. They say we must declare on the butter we export that it is produced from a black animal called the buffalo. This is aimed at putting of Western contumely and killing sales. Then they say export should only from milk that Machine Milked. That is impossible when the milk comes from a farmer who has only one buffalo. We don't need these markets, but all the same we will raise these issues! (Let us feed own people first, than others. (An interview with Dr. Kurien, ECONOMIC TIMES, May, 1998). Incidentally, this statement also shows strategic significance given to exports in Western/Europion nations.

Some of the recent studies highlight the experience of dairy liberalization at village level local city market level with reference to particular milksheds in some of the states line Gujarat, M.P. and Punjab. We may very briefly record the main findings of such study with their important implication from the angle of the future dairy industry scenario and policy implication which we shall discuss in our next final section.

Our micro-level study (Dilip Shah, April, 1996) related to India's one of the most largest dairy plant i.e. Sagar had led us to record the following findings :

"Major Private dairies are as many as 35-40 (Most prominent being Sardar Dairy) accounting for a total procurement share of some 20-25% severely affecting co-operatives procurement in 1994-95 private efforts are more systematic and organized than even before, posing serious threat to co-operative organizations. They concentrate on villages which are close by an and have high density of milk surplus, to the disadvantage of the distant and surplus areas. They mimic co-operatives to create an image of attractive alternatives which would provide better price, market and support in inputs, credit etc. This has weakened the faith and loyalty of member producers to some extent. They have become successful in increasing their market share in liquid milk marketing and have slowly changed the structure of market from a seller's market to buyer's market.

Our another micro-levels study concentrating on liquid milk based dairy co-operative of Gujarat shows the following :

'Private dairy units in the sense of the small scale dairy product manufacturing units increased and their customers associations could promote procurement in the most market close area and best milk production area. During even one year such private dairy societies increased by 42% in Vadodara Milk market also due to private dairies the share of co-operative dairy declined from 70% in 1991 to 62% in 1994-95, member of Private products increased to 35 and types of milk by five types and thus strong competition was for the first experienced by Vadodara dairy (Dilip Shah, Nov. 1996).

One another micro-study of Resurvey of impact of operation flood in M.P. (Hiremath Katarsing and Mergos etc. Feb. 1997) has also recorded modus operandi of private vendors and private dairies in the sample villages of Indore and Ujjan Milkshed (low operating cost, flexible price, liberal credit, monopoly of Mava market, personal and friendly social relations earning the good will layouts of producers are major features of the modus operandi of private dairies which has created a competition for dairy co-operatives.

A study on 'Market oriented Dairying in Punjab' (Amarik Singh, 1999) records in interesting development in the dairy sector of Punjab state. In the wake of delicensing 16 new dairy plants, with handling capacity of around 300,000 liters come up; the state already had a processing capacity of about 35 lakh liters in the co-operative as well as private sector (including MNC). This capacity was marginally in excess of the available surplus of about 30 lakh liters for the organized sector. At that stage, it was felt that the co-operative sector with all the limitations of Government controls and social costs, is placed in a competitive disadvantage against the new entrusts, who were given incentives under the state Industrial policy such as tax exemption for a years apart from capital subsidy, it was thus, speculated by several quarters that the co-operative sector will not be in a position to withstand the onslaught of competition and the days of co-operatives are numbered. However, MILKFED and its members unions with a strength of over three lakh milk producer in the state, coupled with the dedicated workforce stood to challenge and have continued to grow, in terms of rise in procurements, by 15% sales and turnover by 40% (Amrik Singh, Market Oriented Dairying, 1999).

Thus, the post-liberalization period scenario of the Indian organised dairy industry has shown the presence of predictable and unpredictable trend. On the one side, fear expressed by the co-operative sector have not come true on the other side, expected behavior of private dairy units come true and also expected response of developed countries in the international trade has also met with reality.

In the following section, we may briefly underline some major strategic policy actions for the next millennium with an objective of strengthening the base and structure of dairy industry and robust growth of the industry with strong complimentary existence of both co-operatives as well as private sector with fair playing ground for both as organized sector of dairy industry has wide scope for all as it covers just 30% of total production of India.

III STRATEGIC POLICY IMPLICATIONS FOR THE NEXT MILLENNIUM

1. M.M.P.O. was notified in 1992 for orderly growth of dairy sector, however, its provisions could not carried effectively which resulted in 'excess capacity' leading to the threat of inefficient high cost processing calls for the striker and more effective implementation

of MMP. It is not only advocated by the N.D.D.B. (NDDB, 1996, Annual Report). But also realized by the World Bank which has now recommended that "repeal of the MMPO would be premature" as the MMPO provides a bulkwork against over investment, substandard equipment's and unhygienic operation conditions (World Bank, 1998).

2. The major threat and opportunity of dairy industry in India is not from big private companies but mushrooming small units who are capable of avoiding regulations and laws, who adopt all means to destabilize the co-operative units. However, a renewal effort to enforce the provisions of the prevention of food. Adulteration Act of 1954 is extremely required for strict quality control, hygienic standards, adulteration of the milk etc. According to world Bank, (World Bank livestock review, 1996). The state could explore with paying private sector organizations to carry out the inspection and interconnect function.

3. Since the second wave of radical liberalization could not be get in the prediction of entry of new MNC directly could not be successful and, successfully eliminated by the strong performance of co-operative sector, however, it is still a possibility that some of the existing private units would be taken over by MNC in future in which case, not only liquid milk but also the product market also will witness wild competition. Under this situation, World Bank has rightly prescribed that 'All Government incentives for investment in the dairy industry to attract the private dairy operators must be eliminated and all simultaneously, operation flood dairy co-opeatives should be free from all state interventions of the State.

4. Some major facts which have been documented in the study (Ashok Gulati and Raheja, 1990 and World Bank, 1996) are required to be noted to depict the real picture of co-operative dairy sector - major sector in organized dairy sector. The co-operative sector was most protected sector and, in comparison with international standards the processing margin believed to be much higher than international standards must be contained to be able to offer better price to producers. Technical losses and fat and SNF loss be reduced to minimum. To adjust with new situation, co-operative sector has to undertake a large number of policy action at the village level, at the union level and at the federation level which we have prescribed in our more detailed rsearch based studies (Dilip Shah, April 1996, Nov., 1996).

However, it is very admirable that co-operative under NDDB has responded very well. This is evident from the performance reports of federations, maintenance of markets in milk product: continued progress of co-operatives in Non-Gujarat states and remarkable reforms such as 'Total Quality Management' even at the district co-opeative dairy union level.

It is remarkable that India's one of the top ten dairy has recorded a significant progress by qualitative management with better results through optimum utilization of internal resources, removal of waste materials, reduction in the cost and identifying the root causes of the problems (Mehshana District co-operative Milk producers, 1997).

5. In the area of international trade liberalization, it is so obvious that even Gulati States. "But under a free trade scenario, the producers/processors could not compete in the World market and could have been hurt adversely if unrestricted imports were permitted. It would be necessitate continuous monitoring of International markets, and Government will have to act fast to ensure that Indian producers'/processors do not become victims of dumping by dairy giants of the west (Gulati, 1997).

In the area of exports, it is advisable that in order to make sure that import liberalization does not unduly suffers the price of the producers. It is important to simultaneously open up the experts of dairy products. As suggested by experts to begin with say 5% of the last two years rolling production of milk equivalent can be fixed as advanced quota for export of dairy products. Gradually this quota can be abolished! (Gulati, 1997). To promote exports to obtain the benefits of WTO obligation based subsidy removal in E.E.C. and Western countries, our Government must represent our case before WTO to eliminate the will of non-tariff trade barriers in the name of "a draft code of Hygienic practices prepared by the codex committee for Milk and Products which is clearly discriminative to system of milk production in developing countries like India. However, real markets of experts for India include a large 10 million NRI, and country like Shrilanka which imports 90% milk powder and those culturally homogeneous countries which has great potential for Indian Indigenous milk products (Dilip Shah, 1997). We need a powerful export promotion and regulating agency outside the ministry which could be independent and expert in its functioning. N.D.D.B. is one which may be assign the task at least for next 5 years.

6. An expansion of organized dairy sector in liberalized environment and export promotion both depend upon the high rates of growth of production and adequate marketable surplus for exports. It is

quite admirable that as per FAO's latest 'Food out Look' India with Projected output of 74 million tones exceeding 72 million tones world largest milk production of U.S.A. 'INDIA' will be World's top milk producer in 1999-2000 (FAO Food Out Look, 1998-99) and we shall contribute 30% of the World Milk Production by 2020 from the Lower level of just 12% (Amit Sen, Financial Express, April, 1999). Apart from increasing demand for our own people at the rate of 2% income elasticity; we have to increase of production by improving yields per animal which are still not comparable with any advanced country, certainly we are at much lower level in the world when we compare out yield average 5 liter animal buffalo and 1.5 liter of Deshi cow. An intensive programme of production enhancement for cross breeding of cow yielding non-descript cows with high yielding exotic breed and upgrading of non-descript buffalos through selective breeding with high yielding pull breeds must be given very high priority. It is very righly pointed by experts that even most advanced dairies in Gujarat in the name of 'Consumer preference and high preference for buffalo milk as raw material for product manufacturing have under played the importance of cross breed cow milk (A.K. Josehh, 1997). To boost up this cow milk production trend, apart from strictly following two axis pricing policy, the time has come to review the strategy of overall pricing policy also. For innovative productivity/efficiency promotion the producer should at least cover the long run average cost of milk production and provide a reasonable mark up (Katar Singh, V.P. Singh, 1998). For their effective implementation of this policy, studies on cost of production on the line of ICAR studies must be initiated in all major climatic zones in the country and repeated at the regular intervals of at least three years to furnish the objective basis for rational milk price policy on the basis of these studies; dairy co-operative unions and federations should also be provided freedom to get procurement and sales price. Private dairy sector also should asked to follow this basis.

7. Finally, what should be the structure of national livestock policy in next millennium is best explained by R.P. Aneja (R.P. Aneja, 1998 > OED - IRMA workshop on Impact of operation flood, March, 1997). According to Aneja, two major weakness of Indian livestock policy and programmes are 'mistargeting' and 'lack of strategic' levelage. To eliminate these weaknesses better instruments of livestock policy should be so designed that it could not only transform the national and global market pull into an engine for

generating competitive advantage through participation of small producers with ecological sustainability but also achieve quality and economic efficiency. However, this could be only possible if appropriate institutions and mechanisms are designed to support technology, its transfer at producers level where traditional system is followed.

Table-1

Selected Indicators of Growth and Development of India's Industry, 1951-1993

Indicator	Unit	1951	1961	1970	1972	1982	1987	1993
Number of breedable bovines:								
Cows	Million	46.4	51.0	*51.96	53.4	58.7	60.0	65.8
Buffaloes	Million	21.0	24.2	*27.58	28.6	32.3	38.3	39.2
Total	Million	67.4	75.2	*80.66	82.0	91.0	98.3	105.0
Ratio of cows to buffaloes	-	2.21	2.11	1.92	1.87	1.82	1.57	1.68
Total milk production	Million MT	17.4	20.4	@20.7	23.0	35.8	46.1	#66.6
Per capita milk availability	gm/day	132	127	@107	112	136	162	#199
Milk processing throughout	MIllion 1pd	0.2	1.7	NA	3.5	5.0	10.1	16.0
Use of imported SMP	'000 MT	11.9	47.5	19.0	27.0	41.4	37.1	0.7
Use of imported BO/WBO	'000 MT	0.7	-	1.3	3.2	10.6	8.8	Nil

MT. Metric Tonne; 1pd : liter per day
*Estimated by linear interpolation
#Figures are for 1995-96
@Figures are for 1969-70
Source: Annual Reports of Indian Dairy Corporation and Dairy India 1985, 1997.

Tabel-2

Salient Features of operation Flood and Achievements, 1970-96

Features	Operation Flood (OF) Ph		
	OF-I	OF-II	OF-III
Date When started	July 1, 1970	October 1979	2, April 1,1985
Date When concluded	March 31, 1981	March 31, 1895	March 31, 1996
Investment (Rs. crorer)	116.5	277.2	1303.1
No. of Dederatios/Apex Milk Unions set up	10	18	22
No of Milksheds covered	39	136	170
No. of DCSs set up ('000)	13.3	34.5	72.5
No. of Members (lakh)	17.5	36.3	92.63
Average Milk Procurement (mkgpd)	2.56	5.78	10.99
Liquid Milk marketing (IIpd)	27.9	50.1	100.2
Processing Capacity:			
Rural Dairies (IIpd)	35.9		
Metro Dairies (IIpd)	29.0		
Milk Drying Capacity (MTPD)	261.0	507.5	842.0
Technical Inputs:			
No. Of AI centres ('000)	4.9	7.5	16.8
No. Of AI done at end (lahk/year)	8.2	13.3	39.4
Cattle feed capacity ('000 MTPD)	1.7	3.3	4.9

Mkgpd: million kg per day, IIpd; lakh liter per day; MTPD: metric tonne per day Source: (1) Dairy India 1997; (2) Quarterly & Monthly progress Reports on operation flood, NDDB, Anand, 1996.

CONCLUDING OBSERVATIONS:

Thus, our present paper has recorded a remarkable "White Revolution" mainly due to the operation flood project, however, 1991 dairy liberalization policy opened the flood Iates of all which got converted into milk Mafic dominance, milk market deficit over installed capacity and led undue dumping of imports calling for some kind of regulations under MMPO. Though co operatives could survived the first wave of competition successfully, it will have to acquire the cost competitiveness and efficiency as per international standard. Dairy industry of India will have to take adequate measures on the side of high growth of production, productivity, cost minimization and generation of surplus to stabilize the exports of indigenous milk products for which more assured potential exist among NRIs in the world and neighboring Asian Countries.

REFERENCES

1. **Dilp Shah**
   Delicensing Dairy Product Manufacturing - logic and Reality, Financial Express, May 9-10,1992
2. **Dilip Shah** (1996)-I
   Entry of Private Dairy Co-operatives in Mehshana Milkshed of Gujarat.
   - a Rapid Assessment, 1996 (Forthcoming in co-coprativization, liberalization and Dairy Industry in India).
3. **Dilip Shah** (1996)-II
   Private sector Dairies in Vadodara .... Impact and Future strategy for the Vadodara district co-operative union in **Re-discovering co-operation** Vol. III. Institute of Rural Management, Anand (1996).
4. **Katarsingh (1997)**
   Role of Milk Producers' Co-operatives in Market Oriented Dairy Development.... Text of Presentation at the FAO - workshop on Market Oriented Dairying and Role of Producers' organizations and NGO's, December, 1997.
5. Government of India (1976)
   Ministry of Agriculture, National commission on Agriculture, Vol. 1976.
6. **Dilip Shah**
   See for early history
   An Economic Analysis of Co-operative Dairy Industries in Gujarat Somaiya. Bombay, 1982.
7. **Doornboss Martin, Liana Gertsch and Pied Ternal**
   Dairy Aid and Development, SAGE, 199.
8. **Dilip Shah (1993)**
   Reputability, Dependency and equity issues in operation flood project, Journal of Indian school of Political Economy
9. **Wildfred Candler, Nalini Kumar (1998)**
   India's - The Dairy Revolution, 1998.

9-A World Bank Operations Evaluation Department - Production Increase. Is it real?

10. **See for example**
    1. N.C.A.E.R. (1992)
    Impact of Operation Flood Projects.
    2. Dilip Shah (1990)
    Co-operative Dairy Farming Disadvantaged Group Economic Times Part-I & II, 17th, 18th January, 1990.

11. V. Kurien (1997)
    MRF - Annual Conference.... Keynote address (1997).
12. Dilp Shah (1991) Ibid.
13. **Government of India**
    Ministry of agriculture (1992)
    The Milk and Milk product order, June, 1992.
14. **World Bank (1996)**
    India: Livestock sector Review....Report No. 14522, May, 1996.
15. **National Dairy Development Board (1994)**
    Annual Report-1994.
16. **World Bank (1994)**
    An Evaluation of Operation Flood -II.
17. **P.N. Thakur and A.J. Pandya (1997)**
    Dairy Industry of Gujarat... A Profile Gujarat Dairy Industry Vision. IDA. 1997.
18. Govenment of India,
    Ministry of Agricultural, Dept. Of Animal Husbandry and Dairying Annual Report - 1997-98,pp. 23.
19. N.D.D.B.
    N.D.D.B. News bulletin, June, 1999.
20. Hindu business lines, Thursday June, 1999.
21. N.D.D.B.
    Annual Report, 1996-97.
22. **Economic Times**
    HLL puts animal and milk food plants on Sale, 12 June, 1999.
23. Kurein V.
    MRF Ibid, Annual Conference, 1999.
    Competitiveness of Indian Dairy sector
    Paper submitted for the OED/IRMA workshop on Impact of operation flood March, 17, 18, 1997.
25. Harish Damodaran (1999).
26. Dilip Shah
    Dairy Product exports performance and Policy in Post-Liberalization period N.I.R.D. Seminar (1997).
27. World Bank 1998, Ibid.
28. **World Bank, 1996**
    India Livestock Review, 1996, Ibid.
29. **Ashok Gultati, 1990**
30. **World Bank, 1996**
31. **Dilip Shah,**
    April, 1996, Nov. 1996

32. Mehshana District Co operative production union
Annual Report, 1997
33. Gulati, 1999, Ibid.
34. Dilip Shah, 1997, Ibid.
35. FAO
FAO- Outlook - 1998-99
36. Amit Sen, Financial Express, April 1999.
37. A.K. Joseph
Challanges of 2020 for the Gujarat Dairy Industry,
Grjarat Dairy Vision, IDI 1997.
38. Katarsing, V.P. Sing - 1998.
Dairy Development in India - Retrospect and Prospect - Research paper - 15 IRMA , Nov. 1998.
39. R.P. Aneja (1997)
OECD - IRMA Workshop on Input of Operation Flood, March, 1997.

□□□